Building
for Boomers

ABOUT THE AUTHORS

Judy Schriener is a journalist who has been covering the design and construction industry for more than 20 years. Most recently, Ms. Schriener was Managing Online Editor for McGraw-Hill Construction, as well as Editor-in-Chief and a co-founder of Construction.com, the Web presence for McGraw-Hill's Construction Group. Her articles have appeared in *Engineering News-Record (ENR), Architectural Record, GreenSource,* and several regional magazines.

Currently Ms. Schriener, a baby boomer, is a freelance journalist who frequently speaks at trade shows, conferences, seminars, and other meetings, of late mostly about how boomers are affecting and will continue to affect different aspects of life as we know it, as they age.

Ms. Schriener has been honored by many organizations in her journalism career. She may be reached at judy@building-for-boomers.com, on Twitter as www.twitter.com/bldg4boomers, and on the Web at www.building-for-boomers.com.

Mike Kephart is an architect in Colorado who has been licensed in more than 20 states. Last year, Mr. Kephart retired from the practice he founded and led since 1973. During those years KEPHART Community Planning Architecture designed countless residential communities, individual homes, and apartments.

The concept of "housing for later life" has consumed his full attention in recent years both for his own life and the millions of people who face more years beyond 50 than any previous generation. In 2007, Mr. Kephart was honored by the 50+ Council of the National Association of HomeBuilders as their "Icon of the Year."

Mr. Kephart founded Sidekick Homes. Sidekicks are accessory dwelling units (ADUs) that can be placed on single-family lots in the backyards of existing homes or as new construction along with the primary home.

You may contact Mr. Kephart at mike@kephartliving.com and on the Web at www.sidekickhomes.com.

Building for Boomers

Guide to Design and Construction

Judy Schriener
Mike Kephart, AIA

New York   Chicago   San Francisco   Lisbon   London   Madrid   Mexico City
Milan   New Delhi   San Juan   Seoul   Singapore   Sydney   Toronto

The *McGraw·Hill* Companies

Library of Congress Cataloging-in-Publication Data

Schriener, Judy.
  Building for boomers : guide to design and construction / Judy Schriener, Mike Kephart.
    p.   cm.
  Includes bibliographical references and index.
  ISBN 978-0-07-159981-8 (alk. paper)
  1. House construction. 2. Baby boom generation. 3. Architecture and society—
United States. I. Kephart, Mike. II. Title.
  TH4815.4.S37 2010
  728.0973--dc22   2009052953

McGraw-Hill books are available at special quantity discounts to use as premiums and sales promotions, or for use in corporate training programs. To contact a representative please e-mail us at bulksales@mcgraw-hill.com.

Building for Boomers: Guide to Design and Construction

1 2 3 4 5 6 7 8 9 0 DOC/DOC 1 9 8 7 6 5 4 3 2 1 0

ISBN 978- 0-07-159981-8
MHID 0-07-159981-9

The pages within this book were printed on acid-free paper.

Sponsoring Editor
  Joy Bramble

Acquisitions Coordinator
  Michael Mulcahy

Editorial Supervisor
  David E. Fogarty

Project Manager
  Smita Rajan, Glyph International

Copy Editor
  Priyanka Sinha, Glyph International

Proofreader
  Bhavna Gupta and Ragini Pandey,
  Glyph International

Indexer
  Steve Ingle

Production Supervisor
  Richard C. Ruzycka

Composition
  Glyph International

Art Director, Cover
  Jeff Weeks

Other Books in the McGraw-Hill Construction Series

About McGraw-Hill Construction

McGraw-Hill Construction, part of The McGraw-Hill Companies (NYSE: MHP), connects people, projects, and products across the design and construction industry. Backed by the power of Dodge, Sweets, *Engineering News-Record (ENR)*, *Architectural Record, GreenSource, Constructor,* and regional publications, the company provides information, intelligence, tools, applications, and resources to help customers grow their businesses. McGraw-Hill Construction serves more than 1,000,000 customers within the $4.6 trillion global construction community. For more information, visit www.construction.com.

Contents

Acknowledgments   xi

1   Boomers: Who They Are, What They Want, Why You Should Care   1

Introduction   1
Boomer Nation   3
Changes in the Wind   3
Who Boomers Are   6
What Boomers Want   8
Why You Should Care   9
It's All About Lifestyle!   10
References   11

2   Unexpected Challenges   13

The Recession   13
The Hit on Real Estate   14
Back to Work   15
No Going Back to the Future   16
Boomers Are Unprepared   17
Stepping up the Game   19
New Behaviors, New Preferences   19
References   21

3   Neighborhood Types   23

Age-Targeted, Age-Restricted, Age-Qualified   24
Market Size   25
The Big Picture—Large Communities   26
Mega Communities Aren't for Everyone   30
Mixed Generations   35
Infill Neighborhoods and Rehabs   36
Opportunities   36
Challenges Ahead   38
Multistories   39

Urban Downtowns, Suburban City Centers, Traditional Neighborhood Developments, and Transit-Oriented Design   39
Urban Downtowns   40
Suburban City Centers   42
Traditional Neighborhood Developments   43
Transit-Oriented Design   47
University-Affiliated Retirement Communities   49
Existing University-Based Communities   52
New Generation of Communities   53
Guidelines   55
Cohousing   55
Words from the Father of Cohousing   55
Cohousing in Action   58
References   60

4   Aging in Place, Universal Design, Sustainability, and Building Green   63

What's the Difference?   63
Going Green   64
Green Goes Mainstream   64
Building Orientation   65
Energy Conservation   68
Water Conservation   73
More Ways to Go Green   73
Aging in Place and Universal Design   74
Getting Started   77
Elements of Universal Design   80
Summary of Universal Design   92
Bringing It All Together   92
References   94

5   Technology Turns the Tide   95

Not-So-Simple Technology   95
Great Expectations   97

Opportunity Knocks 98
Command and Control 98
Tech Step-by-Step 100
Great Room 100
Kitchen 101
Bed and Bath 101
The Rest of the House 102
At Minimum . . . 102
References 103

6 Single-Family Homes and Townhouses 105

Neighborhood-Driven Similarities and Variations 107
Old Is Now New 107
Resort-Styled Communities Personified by Sun City 110
Traditional Neighborhood Designs 111
Cohousing Community Plans 112
What People Want 112
Preconceptions and Prejudices 113
Home Distinctions 117
Second or Vacation Homes 117
Duplexes 121
Triplexes 121
Townhouses or Row Houses 121
Two-Story Active Adult Homes 122
One-Story Versus Two-Story Homes 123
Accessory Dwelling Units 124
Individual Home Features 130
Ceiling Heights 130
Views 131
Elevators 131
Active Adult Homes Construction Costs 131
Personal Touches 133
Quality Triumphs Over Quantity—Finally! 133
Opting for Quality 135
Smaller Homes 136
Rightsizing 137
References 139

7 Condos and Apartments 141

Apartments or Condominiums: What's the Difference? 144
The New World of Condos and Apartments 146

Age-Restricted, Age-Targeted, or Lifestyle-Targeted? 147
Central City Downtown Locations 148
Considerations for Condo/Apartment Design 150
Location! Location! Location! 150
Amenities 152
Apartment/Condo Elements 155
Storage 158
Parking 158
Wrapping It Up: One Builder's Secrets to Good Condominium Design 159
Urban Housing Models 160
Suburban City Centers 161
Belmar 161
Town Centers 163
References 165

8 The Design Process Step by Step 167

Design Fundamentals 167
Market Research 167
Market Studies 170
Community Vision 172
Community Design 172
Community Amenities 175
Architecture for Homes 181
Home and Community Synergy 183
References 185

9 Design Options 187

Great Room 187
Living Room 187
Dining Room 188
Kitchen 189
Bedrooms 193
Master Bathroom 194
Office 196
Storage 197
Lessons Learned from Experience 198
Give Them Sleep Options 198
Give Them a View Upon Entering 198
Consider a Courtyard 199

*Put the Master Bedroom Suite on
the First Floor   199*

References   200

10 Looking Ahead   201

Trends   201
*Here a Trend, There a Trend, Everywhere
a Trend, Trend   201*
Crystal-Ball Gazing   210
Speculation Based on Forces of Change   211
The Long View   212
References   212

11 Determining and Developing Your Niche   215

Questions to Ask, Steps to Take   216
Asking Questions   216
Taking Action   217
Examples of Potential Niches   221
Location-Based Market Niches   221
Social and Cultural Niches   222
Lifestyle Market Niches   224
Designing for Your Niche   227
Land Planning   227
More Planning for Your Niche   229
Architecture for Your Niche   229
References   230

12 Dos and Don'ts of Dealing with Boomers   233

Call Them This, Don't Call Them That!   233
Generational Roots   234
Coping with Reality   235
Capturing Boomers' Interest   236
Timeline   238
Engagement = Relationships   239
Messaging to the "Who, Me? I'll Never
Get Old!" Generation   240
It's All About Lifestyle   242
Marketing in the New Age   242
Forget One-Way Communication   242
References   245

Appendixes

A Cohousing   247
B LifeCenter Plus   261
C Zoning Rules for Accessory
Dwelling Units, Typical of Others   267

Glossary   269

Index   273

Acknowledgments

"Gratitude is when memory is stored in the heart and not in the mind."
—Musician Lionel Hampton

Nobody accomplishes anything without help, and I have been fortunate to have so many terrific people helping me in so many ways since long before this book was even conceived. Priscilla Wallace, head of the New Marketing Network, clued me in on the whole aging-boomer boom several years ago when she excitedly told me about the work she had begun to market to boomers through her new subsidiary, Boomertising. At the time, she could hardly get anyone to listen to her, and now look at the explosion of awareness of the aging of this incredible "cohort," as they call it. Thank you, Priscilla, for your generosity, enthusiasm, and wisdom.

Thanks to Jan Tuchman and Bill Krizan at *Engineering News-Record* (ENR) for indulging my new passion and letting me cover the construction angle of the boomer phenomenon for ENR. You have been my primary editors for the last 20 years, and I could not have asked for two better editors! Thanks to Andy Wright, also at ENR, for suggesting this topic and me as the author to Joy Bramble, our editor at McGraw-Hill Professional. Norbert Young, president of McGraw-Hill Construction, gave me so much insight into the "real" construction industry over the years. Howard Mager and Howard Stussman hired me at ENR—thank you for changing my life. And didn't we have fun!

Jeff Jenkins, Mark Pursell, and Lakisha Campbell and their colleagues at the National Association of Home Builders were always extremely helpful and gracious. I learned so much at NAHB's conferences and met the cream of the crop in the homebuilding industry and especially builders for boomers and seniors. I am grateful to nearly everyone I met through NAHB, too many to list, but especially to Margaret Wylde, president and CEO of ProMatura Group LLC in Oxford, Mississippi, and Ed Hord, senior principal at Hord Coplan Macht in Baltimore, Maryland. Margaret is a primo researcher in the field of boomers and seniors and from the moment I met her she blew me away, as a researcher, a businesswoman, and a person. Ed, a terrific architect, took the time to answer many of my stupid questions early on, and later I discovered that he was the designer of the active adult apartment complex I moved into, the Evergreens in Columbia, Maryland. I love the place!

I want to thank everyone I interviewed for this book for giving me their valuable time and insight. I truly appreciate it. Ric Johnson, you gave me an awesome interview with your step-by-step infrastructure guide for the chapter on technology. Wow! And, architect and tech guru Paul Doherty, thank you for your ongoing insight and support over the years and especially for the long interviews for this book all the way from Shanghai, China. My early interviews that educated me and grounded me were with Christine Fortenberry of

Fortenberry Construction Services; architect Anne Olson of Olson Architecture; Jim Phelps of Jim Phelps Signature Collection; Jim Chapman of Jim Chapman Communities; Dave Schreiner, formerly of Pulte Homes' Del Webb and currently an industry consultant (and my 50+ guru); Rebecca Stahr of LifeSpring Environs Inc.; Billy Shields of Sunrise Senior Living; and John Rhoad of RMJ Development. I appreciate your time, experience, and insights.

Laurie Meisel and Lareen Strong have guided and helped in invaluable ways as we tried to navigate our way through electronic and social media to let people know about the book. Thank you both so much.

My behind-the-scenes support group was awesome. My Tuesday night dinner buddies at the Evergreens, Mary Ann Kelly, Rita Kerner, Pat Bathras, Dianne Florio, and Cathy Houff, kept me sane and laughing. The whole troupe at the Evergreens, including our property manager and friend, Kathleen Russell—I appreciate you so much. Ginger Evans of Parsons Corp. and Bob Nilsson of Turner Construction showed me by example how to keep things in perspective. Pat Burns of Mortenson Construction in Alexandria, Virginia, gave me ongoing input and encouragement as an engineer, a builder, and a boomer who dreamed of his perfect house. My spiritual advisor Irene Hunter, and my business coach Joyce Reynolds have gotten and kept me on a good track for many years; thank you both for help, guidance, and support that I appreciate beyond what words can express. My free-spirited, friend Michele Delo, artsy friend Susan Shoemaker, and always-there friend Linda Brock always seemed to call or e-mail at just the right time. My mom Darlene Schriener (who, unfortunately, passed away just before this book went to press) and her partner Lloyd Munns contributed to my success and sanity every day with their love and support. And even though my dad, Jack Schriener, has been gone for 15 years, I could hear his encouraging words. Thank you, all.

I can't say enough good things about my agent, Julia Lord, of Julia Lord Literary Management, who is all of the things an agent should be and much more. Thank you, Julia, for everything.

Good friends Gerry Goldberg and Noah Nason, thank you for loaning me money 25 years ago so I could become a writer. Early editors Steve Bergsman, Francie Noyes, and John Craddock, thank you for teaching me how to write for publication. Step-daughter Carey Sweet, thank you for being such a cool person and an awesome writer; I have a lot to try to measure up to.

I feel very blessed to have been able to collaborate on this book with Mike Kephart, a pro in every sense of the word, a true gentleman, and a man with a sense of humor and the patience of Job.

Mike and I both want to thank our editor, Joy Bramble, David Fogarty, and all of the people at McGraw-Hill Professional who guided us rookies through the book-publishing process. The real estate market was flying high when we got the contract for this book, and you gave us the time to start all over when it tanked.

We both also are grateful to Smita Rajan and her team at Glyph International for their professionalism, patience, and guidance from production through printing.

Judy Schriener

I want to offer special thanks to some people who pitched in and helped me write various sections in this book:

Janet Lynne Roots, who edited every word of my first and second drafts before Judy and I were under contract with McGraw-Hill Professional.

David Hall, owner and operator of the best fitness and wellness facility for the over 50 population in the United States.

Charles Durrett, architect and author of two books on the design of cohousing communities. He and his wife Kathryn McCamant are the recognized experts on the subject. Charles literally wrote a complete chapter on cohousing design for us that we spread throughout the text; the full text can be found in App. A.

Sarah Susanka for inspiration in a new way to design homes. Thank you for sharing your "Not So Big" concept and I hope we have carried your ideas to another group of homebuyers not yet aware of the value that can be created by building not too big.

I met many special people during my travels around the country looking for new community design concepts for the 50+ boomers. These people went out of their way to help and spent hours doing just that with no thought to any reimbursement or even mention in this book. They all stand out in my memory so I won't try to rank them in any order of importance. The following list is in alphabetical order.

Blake Gable of Barron Collier spent the better part of a day taking me with him on his tours around the construction site of Ave Maria in Naples, Florida, ending with my meeting the founder and benefactor, Tom Monaghan, who gave me a personal tour of the university campus soon to open.

Tiffany Matteau and Kelly Guarascioin from the Barron Collier offices set everything up for my tour. Without them the day would have been quite different.

David Lepow, president of Centex, Venice Beach, Florida, sat still for an hour of interviewing and then toured the development of Venetian Falls with me to show me just how their design and marketing strategy worked.

Ciji Ware, author of *Rightsizing Your Life*, gave me a copy of her book on boomers prior to its being published and offered her experience to help me with this book.

David Schreiner, former board member of the Del Webb Corporation and president of Sun City Huntley in Illinois. David was not only a valuable resource; he was willing to help us when he could.

Dene Peterson, the holder of the flame that enabled the creation of an entire community called ElderSpirit in Abingdon, Virginia. A former nun, Ms. Peterson kept everyone on track and now lives with the people who built the community with her.

Don Jacobs, a longtime friendly competitor, kindly contributed photos and stories on several of his developments for the 50+ boomers. Don's company, JZMK Architects and Planners, is located in Southern California.

Eric Brown, with Artisan Communities in Phoenix, Arizona, supplied us with all of his professional photography and a very interesting story of why this downtown loft development appealed to the 50+ crowd.

Jerry Gloss, a friend and architect, shared his drawings and photos on the remodeling of his own home. Remodeling is a very big segment of the business of housing for older boomers, as no one wants to move unless they need to.

Thanks to the residents at the Palms of Manasota who invited me into their homes and shared their community with me.

Ralph Spargo, who at the time was with Standard Pacific Homes. Ralph and architect Don Jacobs created one of the few examples in the country of an intergenerational community with a modern flair.

Steve James with DTJ Design in Boulder, Colorado, shared photos and a full description of one of their newest 50+ community designs. Steve has been a colleague and friend for many years.

Others that helped by submitting information and photography of new ideas in housing for the 50+ follow. Many of them went to great effort to respond to my requests and I am forever thankful for their part in this process.

Carson Looney of Looney Ricks Kiss in Memphis; Cheryl O'Brien, C. O'Brien Architects Inc. in Bala Cynwyd, Pennsylvania; Chris Lessard, of Lessard Group Architecture in Vienna, Virginia; Craig Cawrse, Cawrse & Associates in Chagrin Falls, Ohio; David Clinger, land planner in Denver and Evergreen, Colorado; Phil Hove of Hove Design; Dawn Michelle and Don Evans of the Evans Group in Orlando, Florida; Doris Perlman of Possibilities by Design in Denver, Colorado; Jessica Richman with the city of Portland, Oregon, who saw that we were able to reproduce portions of the city code on ADUs; Gary Godden at Godden Sudick in Denver, Colorado; Georgianne Derick of Merchandising East; Kay Green of Kay Green Design in Florida; Chuck Fuhr and Paul Donnelly, builder and developer, respectively, of Madison Lakes in Madison, Georgia; Elizabeth Baker of Rainbow Vision in New Mexico; Tracy Cross of Tracy Cross & Associates in Schaumberg, Illinois; Tracy Lux of Trace Marketing in Sarasota, Florida; and last but certainly not least, Walt Richardson of RNM Architects in Newport Beach, California, a mentor and guide for me throughout my career as an architect.

Mike Kephart

Building
for Boomers

1 Boomers: Who They Are, What They Want, Why You Should Care

"To me, old age is always 15 years older than I am."

—Bernard M. Baruch (1870–1965)

INTRODUCTION

The authors set out to write a book for those who will be designing and building houses and communities for the aging baby boomers who will be reaching their fifties, sixties, seventies, and beyond in the next four decades in numbers too great to ignore.

Baby boomers are the most powerful demographic group our country has ever seen. In 2010 boomers—defined as those born after World War II, from 1946 through 1964—comprise nearly one quarter of the total U.S. population. By 2011, half of all homeowners will be age 50 or over. About 10,000 boomers turn 60 every day. Boomers are still the defining demographic for product design and delivery in the United States—you can see the bulk moving through the body of the population as boomers move through their life cycles.

Those on the leading edge of the wave of boomers began to turn 60 in 2006, while those on the trailing edge are headed toward their late forties. In fact, demographers break boomers down into "leading-edge" boomers—those born between 1946 and 1955—and the "trailing-edge" boomers—those born between 1956 and 1964—with approximately half of the 77 million total boomers in each group. The two boomer segments together will flex their economic muscles like never before as they age and enter this new chapter of their lives. In addition, boomers will continue to set the new, higher standard for succeeding generations in terms of product change and innovation as they have been doing for decades. That provides a great opportunity for the designers, builders, and suppliers who wish to serve this new generation of "mature adults" (though many boomers will tell you they have never aspired to maturity).

Figure 1-1

A baby boomer family on the steps of their post–World War II home in the early 1960s. Note the asbestos siding. (*Photo courtesy of Mike Kephart.*)

This book is for you whether you are an architect, a builder, an engineer, a developer, a remodeler, a do-it-yourselfer, a supplier, a homeowner, or just an interested party. It is for you whether you have been designing or building for "seniors" for many years, or you never have but are guided or forced into it because of the economy or zoning restrictions, or you have the foresight to see that boomers shortly will pretty much *be* the market. The authors want this book to stop you from making costly mistakes and guide you in serving the housing market that boomers are already beginning to shape.

The elements of designing and building for boomers are so numerous that they could fill several books. The authors have tried to compile enough information in one place to get you started on the right track. They have talked with dozens of people; led and sat in on countless seminars and conferences; visited dozens of 50+ housing projects; read hundreds of reports, books, and articles; culled information from scores of statistical studies; and added their own 50+ years of combined experience in the design and construction business. One of the authors is a boomer herself and the other has been designing and building for boomers so long that he knows them better than they know themselves.

The fascinating and exciting aspect of what the authors bring to you is the information that you can use to make the first "footprints in the sand" in the building for boomers housing genre. Nobody knows what to expect. Nobody knows what's going to happen. Nobody has been here before, which is even truer in light of the recent volatile, fluctuating economic conditions the likes of which also have never been seen before. Based on the past, boomers will change housing for future generations. But nobody knows how. You as designers and builders of housing will be working hand in glove with boomers to make those sea changes. Good luck! Building for boomers as they grow older is not going to be like building for seniors as anyone has ever known them. Boomers are, to borrow from the current vernacular, not your parents' seniors.

Here is a look at boomers—who they are, how many of them there are, what they're like, and why you should care.

BOOMER NATION

Baby boomers have been changing the world practically since the first one was born in 1946, in part because there are so many of them.

The baby boomer generation—so named for the "boom" in births after World War II—is generally considered to be those born from 1946 through 1964. The numbers alone are stunning. As of July 1, 2009, boomers numbered 77 million, according to the U.S. Census Bureau. The aging of America is accelerating right now at a record pace. The 55 to 64 age group is the fastest growing age group of this decade, growing nearly 50 percent from 2000 to 2010. By 2011, half of all homeowners in the United States will be 50 or over.

Changes in the Wind

As boomers technically become "seniors," numbers and age ranges don't even begin to tell the whole story about who boomers are, how influential they are, and how they change the world forever as they move through every life stage. The impact that boomers have had on the economy, the environment, the family, the workplace, and the way we live is huge. It started when baby boomers were, well, babies.

By the end of 1953, boomer babies and children numbered more than 30 million in the United States out of a total population of a little over 150 million. Consequently, kids became the center of the business universe. Disposable diapers were invented. The numbers and varieties of baby foods skyrocketed. Disneyland was built. As a block of boomer kids started growing beyond toddlerhood in the 1950s and 1960s, the early surge of sugary kids' cereals appeared: Sugar Corn Pops, Sugar Frosted Flakes, Trix, and Kix. Television erupted with kids' shows: "Howdy Doody," "Sky King," "Lassie," "Mickey Mouse Club." Boomer preteens and teens raced home from school to watch "American Bandstand." All of those TV shows were among the earliest of genres that have morphed and grown into the monstrous selection of shows we have now.

Convenience-oriented boomer teens helped McDonald's grow from one store to the dominant worldwide fast-food chain it is today. Soon pizza delivery, Chinese take-out, and fast-food places abounded, and don't forget TV dinners. The fast food, TV, and leisure environment most boomers grew up in led to a greater need and desire for exercise, and so began the proliferation of jogging and other athletic shoes and well-outfitted gyms. As early boomers became young adults, Disney World was built, with more exhibits about geography and history than Disneyland and more scary or thrilling rides of greater appeal to them.

As boomers began their careers and worked their way up the corporate ladders, they created a social atmosphere in the workplace, turning colleagues into friends and "family." As they married, boomers nudged the divorce rate up to 50 percent and made being single fashionable. Boomer parents didn't want their parents' station wagons but still needed the functionality; voila! SUVs! And now as the youngest boomers lurch past age 45, look at the proliferation of skin creams, hair dyes, tooth whiteners, concealers, and drugs to aid or diminish the ravages of age, pain, fears of losing their beauty, and the paranoia or reality of sexual dysfunction. Boomers have barreled through life demanding innovation, change, convenience, and creature comforts, and they are used to getting all of that. Mike Lynn from Boomertising, a Chicago-based marketing firm specializing in targeting baby boomers, calls this drive "vanicessity," a cross between vanity and necessity. Boomers have wanted it all and have been used to getting it. Their parents, after surviving the Great Depression and World War II, strove to provide them with everything they had had to do without.

Paradoxically, boomers have always been concerned with bigger issues than themselves. They've tried to garner fairness for all, stop war through protest after protest, and save our planet. In the 1960s, when the oldest boomers were in their teens and early twenties, they spawned the protest generation, convinced they would end war forever. They also created communes, a new form of community started with the best of intentions. War continues today and communes are a distant memory, although, as you will see in Chap. 3, the concept of cohousing, which is sort of communes for the twenty-first century, is reemerging.

Boomers created a public awareness of the environment. Earth Day, environmental impact surveys, wildlife preservation, antilitter campaigns, and energy conservation all began in earnest under the influence of boomers.

Suburbs and Antidotes

For better or worse, boomers' greatest legacy may be the suburbs. Boomers fueled the inception of the massive suburban growth after World War II by their very births (see Fig. 1-2). Suburbia was born as Levittown, New York, the first planned community of mass-produced homes, and blossomed between 1947 and 1951.

Both of the authors grew up in these early suburban houses. Kephart, his brother, and his parents shared a two-bedroom, one-bath house of 900 square feet. The asbestos-siding-clad Cape Cod came with an unfinished basement that leaked too much to think of finishing and an unfinished attic his brother and he later shared as their bedroom. Schriener, an only child, moved as an infant with her parents from the second-floor apartment of a multistory duplex into a small, stark-looking home in a brand new neighborhood of starter homes. With each of the first three moves in Suburbia over the next 15 years, the Schrieners got farther away from the heart of the city, which her father commuted to and from each day.

Boomers as adults continued to power suburban growth into the institution it is today. Mass-produced affordable houses for the working middle class, coupled with the ever-growing variety of inexpensive mass-produced cars, changed forever the way Americans live.

Raised in the realm of privacy in the suburbs, many of today's boomers have had enough of being cut off from others. One response is "new urbanism." The new urbanists have eloquently articulated the problems boomers saw with life in the suburbs, and

Figure 1-2

Aerial view of a typical suburb. (*Photo by Mike Kephart.*)

this is detailed in James Howard Kunstler's book *The Geography of Nowhere: The Rise and Decline of America's Man-Made Landscape*. He writes:

> The great suburban build-out is over. It was wonderful for business in the short term, and a disaster for our civilization when the short term expired. We shall have to live with its consequences for a long time. The chief consequence is that the living arrangement most Americans think of as "normal" is bankrupting us both personally and at every level of government.... Now that we have built the sprawling system of far-flung houses, offices and discount marts connected by freeways, we can't afford to live in it.

Another model of the new culture of living being developed by and for boomers is cohousing. Homes each have their own intact, complete floor plans, but they are clustered around common or shared facilities and amenities intended to bring people together on a daily basis.

At its finest, residents in a cohousing community watch out for each other, often dividing up tasks related to keeping all of its residents happy and healthy. Neighbors support each other in childcare, through injury and illness, and, finally, with aging. Cohousing may exist in an enclave or just be part of a neighborhood, but its chief characteristic is the commitment of neighbors to fellow neighbors who have all chosen to live independently through community. Maybe the great desire of people will no longer be for privacy but for community in the broadest sense: a practical and caring support system for young and old alike. See Chap. 3 for more information on cohousing.

Who Boomers Are

Boomers are not a homogeneous group. First, a cautionary note: "boomer" is not another word for "senior" and it never will be. That especially holds true for anyone who wants a boomer to buy anything from them. Seniors as we have known them for decades refers to a relatively insignificant consumer group in marketing terms, heretofore considered a fairly homogeneous group. Seniors' lifestyles and attitudes have been more similar than varied. Seniors historically were isolated; many of them didn't travel until later in life. A sizeable number of seniors came from the scarcity of the Great Depression and its aftermath, were grateful for their jobs, and stayed with the same employer for many years until they retired. When they did retire, they drifted away, moving to huge Sun City– or Leisure World–type age-segregated communities away from cities, towns, and civilization. Seniors' worlds have been much smaller all their lives than the worlds of the bulk of boomers.

The generation between the Great Depression era and boomers, known as the *silents,* or the *Eisenhower generation*, are the peacemakers of the world. They have continued to populate the communities and facilities for the aging, such as assisted living as well as large active adult resort-styled communities. The transition from the silents to the boomers in the marketplace is well under way already but the boomers will constitute 100 percent of the market for 50+ housing in a few short years.

Leading-Edge Boomers, Trailing-Edge boomers

Within the boomer cohort (another word for generation) are leading-edge baby boomers and trailing-edge baby boomers or late boomers (which you don't want to confuse with late bloomers).

Leading-edge boomers were born between 1946 and 1955. A little more than half of all boomers are considered leading-edge boomers. Trailing-edge boomers were born between 1956 and 1964 and claim a little less than half of all boomers. The two segments together make up what is considered the boomer generation, but the life experiences of the two groups are quite dissimilar.

The Vietnam War pretty much defined the leading-edge boomers, whereas it was all but over by the time trailing-edge boomers were old enough to care about such things. The leading-edge boomers have forged the path for both segments over the years. They have also gotten most of the attention, much as the first-born child winds up in thousands of pictures in the scrapbook but the second appears in a fraction of that number.

Leading-edge boomers began hitting 64 in 2010, just on the verge of Medicare, Social Security, and seniors' discounts everywhere that offer them. However, don't you dare call them old. They're not, just ask them.

Roots

Boomers, even as they get to be "of a certain age," come from an entirely different body of experiences compared to the World War II generation and the silents. They will not be like any seniors the world has known. Boomers grew up, by and large, in abundance.

They traveled early in life, expanding their worlds beyond their home towns. The world was open to them. They questioned everything and were allowed to do it or did it anyway. That evolved into the growth of powerful mass protests to push for changes in the status quo. With boomers' backgrounds being much more expansive and diverse than those of their parents, they grew in different directions much earlier than their parents, and ended up taking many different paths in adulthood. Compared to previous generations, boomers are more racially and ethnically diverse and are better educated. They married later, concentrated on their careers more, had children later, and were the first women to gain significant stature and power in the workplace.

Boomers also are different from their predecessors in other ways. Unlike previous seniors who worked hard for decades and looked forward to retirement as a reward, filled with leisure activities and hobbies, for boomers "retirement" as such is not particularly something they strive for. They don't see themselves retreating from the world as they reach their sixth, seventh, and eighth decades. Some will formally retire, of course, but all surveys the authors have seen and the boomers they have talked to about this subject indicate that by and large boomers expect to remain active and involved in the mainstream of life. How that manifests will vary from person to person. Some will keep working until they draw their last breath. Others will work part time. Many will end one career and start another, possibly going back to school or turning a passion or hobby into a job or a business. Many will no longer commute but will set up a home office. Some will "work" in charitable pursuits or create some combination of volunteering and working for pay. Some will work full or part time for a portion of time each year and travel the world or visit friends and family the rest of the year.

Gerontologist Ken Dychtwald said that people over 50 will be going through a paradigm shift—from linear to cyclic. People used to live life in a linear way: education, work, leisure. "You learned, you worked, you rested, you died," he said ... and you did each of those things once. Boomers will change all that, repeating the education-work-leisure elements throughout their lives, not always in the same order. He said "I think the biggest transformative change that's going to come into our lives as a result of greater longevity will not be that we'll all be old longer, but rather [that] we'll have a new paradigm for the way we'll live our lives. Let's call it a cyclic life paradigm in which it will become normal for people to go back to school at 40 and 60 and 80. It will become normal for people who've been knocked down by a failure in business or in romance or in health to make a comeback. It will become normal for people to ignite talents that perhaps they put to sleep 20 or 30 years ago. It will become normal for us to imagine who we might become at 50 or 70 or 100."

Currently, boomers make up nearly 40 percent of the workforce. The number of workers over the age of 65 will rise 80 percent between 2006 and 2016. Many of those will be forced to work longer than they originally intended for one or more of several reasons. Maybe they have not saved enough money to stop working, or they have unexpected expenses related to health or support of children or parents, or their jobs or 401(k)s and IRAs were hit hard by the recent economic downturn. Retirement assets in IRAs and 401(k)s in September 2007 were $8.8 trillion. In November 2008 they were $5.8 trillion. It will take awhile to get all or most of that back. In fact, older boomers may have to work nearly a decade longer than they had planned if they want to get back what they lost.

The variations in lifestyles of boomers will be more numerous than those of their parents, which will present a great challenge for home designers and builders who for decades got away with offering a handful of very structured floor plans per community. They—you—will have to figure out how to accommodate their desires and still remain profitable.

Boomers think of themselves as forever young. As boomers get to the fringes of what used to be considered senior or elderly, if you're smart you will not ever call a baby boomer a senior, let alone elderly, even when they're 90. Boomers have an image of themselves as young, vital, active, and healthy, and they think they always will be. As one TV commercial for hair color for men says, this is the generation that vowed never to get old. Priscilla Wallace, who founded Boomertising, a division of Chicago-based New Marketing Network, both of which she heads, likes to cite a study she saw in which boomers consider that they will think of themselves as old 3 years after they're statistically dead. See more about how to deal with boomers and what not to do in Chap. 12.

What Boomers Want

The main thing to note is that boomers—as one big group—don't want any *one* thing. This is really the first time that members of an entire generation have not been in lock-step with each other when they've gotten to what some have called their golden years. Boomers are the free generation! They have been freer than any other before them to leave the nest and explore parts unknown both physically and mentally. Therefore, they have scattered in more directions than we can count. What individual boomers want is directly tied to what stage of life they're in and what circumstances they have to deal with. Age is relatively inconsequential.

There are many boomer parents who are grandparents and even great-grandparents by the time they're 60. On the other end of the spectrum, there are also 60-year-old boomers who are parents of infants. Some of those same new parents also may be grandparents at the same time. A single boomer with a child or two at home is likely to want a house more similar to what a young couple would like than a single boomer or couple their same age with no children or parents to accommodate or take care of at home.

Take the layout of the bedrooms in the home, for example. The first question to ask a boomer is, "What is your lifestyle?" A single person or couple with young children or grandchildren they frequently babysit will probably want at least one of the additional bedrooms close to their master bedroom suite. A single person or couple that has no dependents at home but wants room for guests or grown children when they come to visit will probably want the bedrooms on the other side of the house from their master bedroom suite. That isn't written in stone, however. One of the best-kept secrets is the increasing number of couples that are opting for a bedroom close to the master suite to give them a place to get away from a snoring or ill partner, or just because they want to be close but in separate quarters.

Social Aspects

Just 3 percent of active adult homebuyers surveyed said the social environment in their current active adult community did not meet their expectations; 94 percent said the social environment met, exceeded, or very much exceeded their expectations, with 34 percent putting themselves in the "very much exceeded" category, according to ProMatura Group's 2009 study of active adult homebuyers.

You need to go room by room and site by site to see where your boomer prospect's lifestyle influences their choices. Keep reading. That's what this book is all about!

On the flip side of that coin, you need to look at the homes you are offering and figure out what lifestyles would fit and tailor your marketing materials to prospects with those lifestyles. See Chap.12 for more on the dos and don'ts of dealing with boomers.

Why You Should Care

"What you ARE now is where you WERE when . . ." your values were being pro-grammed, which is up to age 20, says popular corporate training speaker Morris Massey. What major occurrences occurred during the baby boomers' early lifetime, especially those leading-edge boomers born between 1946 and 1955 who will lead the way for the rest of their generation as they always have? In other words, what defining events did they witness while growing up that helped to form their perceptions of the world and shape their attitudes and behaviors:

- Rock and Roll (which parents hated) was born in the early 1950s, got more popular throughout the decade, and dominated in the 1960s.
- The space age began in 1957 with Sputnik I, but it was not the dominant United States that launched it but rather the Russians, with whom the United States was in what was known as the Cold War.
- The nation elected John F. Kennedy as its president in 1960, and the time that he and his young family occupied the White House was widely known as the "Camelot" years.
- The Cuban Missile Crisis put the United States on the verge of war with Russia and the threat of nuclear war became a true possibility.
- Kennedy was assassinated in 1963. All three major television networks broadcast an unprecedented (before or since) four continuous days of coverage of the death of a president. Boomer youngsters were able to watch nearly every moment of the after-math, the procession to the church, the funeral mass, and the burial on TV.
- The war in Vietnam that quietly began in the late 1950s, escalated in the mid-1960s, and, despite protests that frequently turned violent, dragged on for more than another decade. Returning soldiers were scorned or ignored by a society that hated and rebelled against the war.
- "The pill" was developed, which gave women more freedom over their bodies and helped to launch the sexual revolution of the 1960s and 1970s.
- The Civil Rights Act of 1964 marked the beginning of a new era in race relations in the United States. Four years later, Martin Luther King was assassinated.

- The Watergate break-in led to the tearful resignation of President Richard M. Nixon in 1974.
- Credit cards became popular; BankAmericard (now Visa) was launched in 1958, MasterCharge (now MasterCard) in 1966.

So, those boomers have gotten spoiled by their parents and been catered to by society. They've questioned authority, rebelled against the status quo, and protested the atrocities of war and the unfairness of prejudice. They've been shocked by the assassination of one U.S. president and the resignation of another. They are disappointed that the world did not change with Woodstock and that wars continue to erupt around the world. Boomers may not have invented the green movement but they have been responsible for a large part of its growth.

Okay, so you've got a spoiled, rebellious, cynical, freedom-loving group on your hands who aren't afraid to cause trouble to make their point. What in the world would you want to do business with *them* for?

Several factors make the influx of boomers into their senior years significant for home designers and builders:

- There are going to be so many aging boomers that even a slice of the boomer pie will leave plenty for everyone. If more than 65 million boomers in the United States will be moving through their sixties and above over the next couple of decades, and even 1 in 10 decides to move, that leaves 650,000 boomer prospects alone! The generation following doesn't begin to live up to those numbers, so why not go for a smaller slice of the biggest pie?
- Boomers as they get older are not going to settle for the same housing styles that previous generations had. They will be vocal and forceful, as they have been all of their lives, pushing for more innovative designs, more flexibility to adapt to their individual lifestyles, and more technology-friendly homes. They will create change for themselves and also for all Americans, as has happened throughout their lives. If you are not on the boomer train, you'll be left at the station wondering where everybody went.
- The opportunity has never been greater to be a part of an exciting, vibrant housing innovation boom. If you love what you do, don't you want to be involved, participate, and contribute?

IT'S ALL ABOUT LIFESTYLE!

Now, after raising their children and creating the age of technology, boomers are determined to design better places to live out the next stage of their lives. The variable is what that next stage will be. Boomers will forge their own patterns, and they are likely to be far different from those of their parents and grandparents. Boomers' requirements and desires are more varied and diverse than those of previous generations, and that will be reflected in the homes they seek.

Among the parents of baby boomers, only 31 percent ever chose to move from their homes and neighborhoods unless forced to do so by disease or frailty. For boomers, it's double that. Of the 77 million baby boomers, 30–40 percent of them—23–31 million individuals—are likely to relocate for retirement or for the next phase of their life, according to the 2010 Baby Boomer Study by Del Webb. Other, more recent studies, generally speaking, concur. It is worth noting that in an earlier Del Webb study, just 7 percent of them are likely to consider an active adult community. So clearly they will be seeking alternatives that probably do not exist today.

One important thing to remember is that boomers' influence is just beginning and will go on for another decade and a half. Their numbers will be formidable the entire time. In 2024, when the youngest boomers turn 60, the U.S. Census Bureau projects that boomers will number over 67 million. The market will be there. Boomers do change things. Your opportunity lies in meeting the challenge of inventing new ideas and new community forms for boomers now who will influence what kind of homes people live in for a very long time.

REFERENCES

1. *Age Beat, The Newsletter of the Journalists Exchange on Aging*, No. 24, "Special Report," Summer 2007, American Society on Aging, 2007.
2. Del Webb, Baby Boomer Study, 2003.
3. Del Webb, Baby Boomer Study, 2010.
4. Dychtwald, Ken, "The Cyclic Life Paradigm," video clip available at www.dychtwald.com/highlights. Accessed September, 2009.
5. Frey, William H., "Mapping the Growth of Older America: Seniors and Boomers in the Early 21st Century," *Living Cities Census Series*, The Brookings Institution, May 2007.
6. Gandel, Stephen, "Why Boomers Can't Quit," *Time,* May 25, 2009, p. 46.
7. Green, Brent, *Marketing to Leading-Edge Baby Boomers,* Paramount Market Publishing Inc., Ithaca, New York, 2005.
8. "Housing for the 55+ Market: Trends and Insights on Boomers and Beyond," Report, National Association of Home Builders and MetLife Mature Market Institute, April 2009.
9. Kunstler, James H., *The Geography of Nowhere: The Rise and Decline of America's Man-Made Landscape,* Free Press (imprint of Simon and Schuster), New York, 1994.
10. Wylde, Margaret, *Boomers on the Horizon: Housing Preferences of the 55+ Market*, BuilderBooks (National Association of Home Builders), Washington, D.C., 2002.
11. Wylde, Margaret A., *Right House, Right Place, Right Time: Community and Lifestyle Preferences of the 45+ Housing Market*, BuilderBooks (National Association of Home Builders), Washington, D.C., 2008.
12. Wylde, Margaret A., "50+ New Home Buyers: Why, Where, What and When," Research study presentation at International Builders' Show, January 21, 2009. (President of ProMatura Group, LLC, Oxford, Miss.).
13. U.S. Census Bureau, "An Older and More Diverse Nation by Midcentury," press release on report released on August 14, 2008, available at http://www.census.gov/Press-Release/www/releases/archives/population/012496.html. Accessed August, 2008.
14. More material in this chapter came from various charts, reports and press releases from the U.S. Census Bureau.

2 Unexpected Challenges

"It is not the strongest of the species that survives, nor the most intelligent but the one most responsive to change. . . ."

—Charles Darwin

THE RECESSION

The biggest shock to hit in recent history that will impact the housing market in general—and the preferences of age 50+ homeowners in particular—for years and years to come was (no surprise) the recession. It consumed much of the last half of the first decade of the twenty-first century. Regardless of when "the recession" officially began, housing began to soften in late 2005 and early 2006. Three years later, housing prices were plunging, sales were stalling, inventory was stacking up, joblessness was skyrocketing, and foreclosures were soaring. It was a full-blown recession and it hit hard. Even after economists declared it officially over, the high unemployment rate, the massive foreclosures, and slowness of corporations to gear up again kept the housing market off balance.

The housing sector started it all but the decline of consumer confidence was the lit match that ignited the gasoline that created the economic meltdown in 2008 and 2009. It was a vicious cycle. Slight declines led to media reports, which led to heightened awareness, which led to fear, which led to pullbacks in spending, which led to more alarming media reports, which led to greater widespread fear, which led to even more frozen spending, which led to layoffs, which led to scary media reports about joblessness and foreclosures, which led to people getting even more fearful and hoarding their money, which led to. . . . Well, you get the idea.

Consumer confidence regarding retirement eroded between 2008 and 2009 to an alarmingly low level. In 2007, 27 percent of workers were very confident they would have enough money for a comfortable retirement. By 2009, that had plummeted to just 13 percent, according to 2007 and 2009 Retirement Confidence Studies conducted by the Employee Benefit Research Institute (EBRI). Even those who had already retired had lost faith in their own financial security, the level of confidence sinking from 41 percent in 2007 to 20 percent in 2009.

The Hit on Real Estate

Regardless of how much of the recession was exacerbated by the panic that people felt and the media fueled, the results were all too real and all too devastating. The stock market plunged nearly 45 percent between its October 2007 high and February 2009. It hit the 50+ crowd hardest. They suffered losses of $5 trillion out of the total $7 trillion lost in that time. Some 50+ investors took their losses and pulled out of the stock market, never to return. Some 50+ investors are back in the stock market but not at the same level of commitment as before. Some are all-in and hoping their retirement funds outlast their retirement. Their retirement savings, including their house, which they thought was a slam-dunk as a secure investment, were decimated to the point that many have postponed retirement or gone back to work when they hadn't intended to. A lot of people who thought they were comfortable for life now have to live as spendthrifts.

The stock market is one thing. People know the risks up front. But real estate has for decade after decade been the secure investment. For most people, their house is their most valuable financial asset. For many boomers, the equity in their house will mean the difference between struggling and living comfortably in their later years. The value of one's house always went up; it was just a matter of how much, how fast. Then it all crashed, taking out gobs of equity with it. It hit every sector of residential real estate. People who bought in active adult communities had an average of $218,000 left after selling their home and paying off the debt in January 2007, but took a nearly 55 percent hit by January 2009, when they had just $99,000 left after selling their home and paying off the debt, according to a 2009 study of active adult homebuyers by ProMatura

Figure 2-1

Many boomers thought they would be able to travel and enjoy life indefinitely. (*Photo courtesy of Bobby Lipp.*)

Group, Oxford, Mississippi. Even those people who didn't sell their homes during the worst of the down time stand little chance of recovering all of the value that was lost in the recession, at least for many years.

Boomers who have been living in large houses may have the biggest challenge of all. In order to downsize or move, they have to have someone to sell their house to. Big houses are getting to be less and less fashionable, and after the recession not as many people can afford them anyway. To get free of their large houses, boomers may have to take less for them than they anticipated and consequently buy another house that's either smaller or in a less desirable location than they would like.

Back to Work

It is fortunate that boomers like to work because they may have to. The trend toward wanting to work at least part time started before many boomers hit 60. Even in 2007 when the economy was strong, an AARP study of workers age 45 to 74 showed that 7 out of 10 plan to continue working well into what is traditionally thought of as retirement years. A 2009 AARP study of workers age 45 and older found that 22 percent of respondents between the ages of 45 and 54 and 27 percent of respondents between the ages of 55 and 64 had postponed plans to retire. More than 15 percent of respondents between the ages of 45 and 64 reported that they had taken a cut in pay. A study by the Pew Research Center's Social & Demographic Trends Project released in September 2009 showed that the recession had cut even deeper. Workers between the ages of 50 and 61—the heart of the leading-edge boomers—said the economy could force them to delay their retirement. In EBRI's 2009 Retirement Confidence Survey, 28 percent of workers said the previous year had changed the age at which they expect to retire, with 89 percent of those saying that age has gone up. They cited the poor economy and their need to make up for losses in the stock market as their primary reasons for thinking they would retire later. Very few said they had changed their expected retirement age because they wanted to work longer.

The good news in all of this is twofold. Boomers want to work and, for the most part, boomers are able to work.

However, boomers' denial before the recession, the fact that their life expectancy is longer and there's a good chance many of them will outlive their money, and the hit they took during the recession all add up to a very different lifestyle than a lot of boomers ever imagined in their wildest nightmares. They thought they were set for life (See Fig. 2–1).

Waiting for What May Never Come

When author Kephart started doing the research for this book in 2006 and 2007, he was always left with the feeling that the reported wealth of the baby boomers was overblown and perhaps more myth than reality. Most boomers still expected to be able to buy a new home for $100,000. What kind of house does that buy today? There was talk of inheriting that wealth from their Eisenhower-generation parents, but that didn't work out for Kephart and his brother.

Their dad passed away before their mom by a number of years but not before putting a real strain on their savings and his pension with the care he required at the end. He suffered with emphysema after smoking his entire life. Heart disease and other maladies complicated his condition until one or more of them overcame his remaining strength. Their mom smoked as well and had a heart condition and amazingly high blood pressure, which Kephart inherited. Her healthcare and housing as her health deteriorated took all she had left financially. The Kephart brothers inherited a house with nearly as much debt as value and split the proceeds. Author Kephart came away with his son's favorite cuckoo clock that was and is still a family heirloom that all the kids love. His wife got a brand new Kitchen-Aid mixer that his mother always wanted her to have. That gets used for Thanksgiving Day meal preparation and maybe for Passover Seder.

His wife's parents, God bless them both, lived as if they were wealthy and the entire family thought they were until they died nearly flat broke. No one suspected. They had a condominium worth less than $30,000, a Ford Taurus worth $3,000 or $4,000, and a bunch of worthless costume jewelry. Little remained after it was all over.

Kephart realizes that his family is not necessarily typical but what is fairly typical is the longer life span the parents of boomers are now enjoying. Kephart suspects their enjoyment is somehow tied in to spending any possible inheritance that their children may have expected prior to the current recession. He also suspects that this recession is just the wake-up call we all needed before it got any worse.

So where is that wealth the boomers were to inherit? Maybe it is still to materialize but he suspects it is time for boomers to buckle down and get back to work. If a windfall comes, then everybody can celebrate, but to live our lives as if that windfall is on the way will only slow us down at the precise time we need to get with it.

No Going Back to the Future

If you are waiting for things to get back to where they used to be, you're in for a long wait! Many futurists say we'll never see that kind of free-wheeling real estate market again, nor will much else be the same either. Futurist Andrew Zolli said, "This isn't a recession. It's a reset!"

Zolli sees a shift coming in our society in the areas of work and family that will affect every area of our personal and business lives, including homebuilding. As half of boomers retire and the other half keep working, boomers will find that the market and employers want them, he said. More than 90 percent of the growth in the U.S. labor force from 2006 to 2016 will be in the 55+ age group, according to one estimate cited by the Pew Research Center. This should bring about a better financial situation for boomers in their later years and also will make the kinds of homes they will want when they are older very different from those of previous generations in terms of locations, floor plans, and amenities. (See Chaps. 6 to 8 for more information.)

Today we send our grandparents away, Zolli said, but retirees will be leaving the Sunbelt as they run out of money as a result of the recession and living longer than they had anticipated. They will be going "home" to their children. They will be buying adjacent properties, living in granny flats—small cottages on the same property, also known as accessory dwelling units (ADUs)—at their children's homes, or moving in with their children in their own wing of the house. These retirees who will be returning to live with their kids ultimately will include boomers. "People born in 1970 or later are all but guaranteed to take care of their parents," he said.

Zolli said he sees a "green collar" economy coming after the "reset." A carbon-constrained economy will put pressure on people to deal with climate change, emissions, and other environmental issues, as well as bring about the return of the laws of supply and demand. "Be prepared for $20-per-gallon gas," he said. That, as we know from $4-per-gallon gas, changes how people live and where they want to live. Zolli predicts a green-focused recovery that will impact the transportation system and bring about a nearly 50 percent growth of the concept of LOHAS (lifestyles of health and sustainability) over the next decade. (See Chap. 3 for more information on LOHAS and Chap. 4 for more information on green issues.)

Boomers Are Unprepared

Nearly one-third of workers age 35 and older said in the 2009 EBRI survey that they were not currently saving money for retirement. Nearly three-quarters of workers in every age group said they expect to supplement their retirement by working for pay after they retire. That's up from two-thirds in the 2007 survey.

The EBRI study also consistently has found over the years that far fewer retirees end up working for pay than expected to. For example, despite nearly two-thirds anticipating that they would work for pay in 2007, only about one-third reported in 2009 that they actually worked for pay sometime during their retirement. Although over three-quarters of those retirees who did work for pay said they did it because they enjoyed it or wanted to stay involved in the workplace, over half also cited financial reasons for working sometime during retirement.

The median age that workers said they expect to retire has been slowly creeping up and is now 65, according to the 2009 EBRI study. However, the study also revealed that the reality may be different. Nearly half (47 percent) of retirees in the 2009 study said they left the workforce before they intended to, with just 10 percent claiming all positive reasons for their early departure. They cited factors such as their own or a spouse's disability or other health problems, losing a job, other work-related reasons, or outdated skills as contributing to their leaving the workforce earlier than planned.

In the 2009 EBRI study, between 6 and 13 percent of workers in three different age ranges from age 35 and over said they expect never to retire.

In order to retire comfortably, experts say that people need to either cut down drastically on what they spend or generate enough income to supplement Social Security, pensions, and income from savings up to a level that's acceptable to them, say 70 percent of preretirement income. Many unknowns exist that make estimating those numbers a little like reading a crystal ball with a blindfold on. People can't accurately foretell how long they will live, what their healthcare and other expenses will be, or what their rate of return on investments will be, any one of which could throw a huge monkey wrench into what they thought would be a financially smooth retirement.

Even before the recession, studies warned that while many baby boomers were on track to enjoy a comfortable retirement, many were not. Industry consultant David Schreiner, when he was vice president of active adult development for Pulte Homes, cautioned builders eager to tap the 55+ market to "make your homes affordable" over and over when other builders were dreaming of the big, expensive houses and lavish upgrades the wealthy boomers would want. Given the impact of the recession, more boomers than ever are at risk of living longer than their retirement income and savings can support them. Studies have been predicting since before the recession that the gap will be wider than ever between the haves and have-nots when boomers get to retirement age.

The GI generation and the silent generation, both of which came from scarcity, were good about saving money. Boomers and Gen Xers, who came from abundance, may be fooling themselves when it comes to preparing for their golden years, whether they plan to officially retire or not. In the 2009 EBRI study, 30 percent of Gen X workers age 35 to 44 said they needed to accumulate less than $250,000 for retirement; 40 percent of boomer workers age 45 to 54 agreed. Just 9 percent in each of the two age groups said they would need $1.5 million or more in retirement.

Some boomers will keep working well into their seventies and eighties. Some will cut back on their spending. Some will recover most of what they lost through wise investing. Some will move in with family or take family in to save money. Some will adjust well. Some will adjust because they have no choice.

Fortunately, baby boomers are all about change. They create change wherever they go. They adapt to it more easily than any generation before them. They embrace it. And they are not easily intimidated. So boomers will adapt to their changing economic situation. They will see it as a challenge, one more way they can positively influence the world, and one more thing to compete on.

In an article on "The Incredible Shrinking Boomer Economy" in *BusinessWeek* magazine, author David Welch sums up boomers' attitudes postrecession:

The trick will be finding a way to fulfill the needs and wants of a generation that is used to being catered to—but is now on a budget. Timothy Malefyt, an anthropologist who studies consumer trends for the ad agency BBDO New York, argues that boomers, having ridden a wave of technological change, are highly adaptable and well versed in problem solving. (Or at least they see themselves as such.) Already, he says, they are making a virtue of value shopping, once viewed by this group as hopelessly déclassé. For many boomers it's no longer about keeping up with the Joneses, it's about outthinking them. "If you make boomers feel they've failed, you'll lose them," Malefyt says. "They want to feel they've outsmarted the system or their circumstances."

Stepping up the Game

The consequences of the recession have changed homebuyers. Therefore, home design and building must change too. Possibly the biggest relatively permanent impact the recession will end up having on the housing market for boomers is simply that homebuilders will have to step up their game to satisfy a more discerning market. Boomers will be looking to get the maximum advantages, features, and value from their homes. Schreiner, the industry consultant and former vice president at Pulte Homes, said: "We deluded ourselves in the business. We said people are okay with a lack of storage or two eating spaces. We're now in a market where people are buying homes to live in. In Florida, look at how many people bought them to move into: less than 30 percent. They bought them for speculation or for a second home. The economic challenges have changed people. A lot of the compromises in livability that occurred in the mid-2000s look ludicrous to a consumer who's concerned with value and real-world issues."

NEW BEHAVIORS, NEW PREFERENCES

Two or three years ago, homeowners were blissfully planning to move to nicer homes or remodel their current ones. They were willing to give financial help to their children, grandchildren, or parents to find a place of their own, or expand their own homes to accommodate an extended family. They were looking forward to replacing worn carpets and curtains; they wanted to repaint their places and spiff up both the outside and inside.

Not anymore. Homeowners have gotten scared and pulled back in all areas, according to three different studies of baby boomers conducted late in 2008 by AARP and released in February 2009.

People are postponing or cutting back on home modifications out of necessity or out of fear of conditions getting worse, whether specifically their own or in general.

Many people indicated they had been planning to move but in light of the difficult economy were now opting to stay put indefinitely or at least sit tight for the moment. The percentages varied among the studies but the trends were consistent. In addition, according to the report, "some are not confident overall about having the money they need to live comfortably in retirement."

Some of the comments from the focus groups of 45- to 64-year-olds in the AARP report are:

"I've held off on making improvements to the house because I don't know if I would get the value back." "With the housing market the way it is and the prices of homes we've decided not to sell at this time." "We don't have major things to do; we're just not doing things we would have liked to do a long time ago." "I have 2 grown sons in California and eventually we'd like to move back to California if we can ever afford it. . . . It's less likely now. (How is that affecting your retirement plans?) Everything's on hold."

Eventually the recession will be a distant memory, but the effects of it may never go away for this generation.

Just how many people will actually move now, versus before the housing market tanked is also up for debate. Two of the AARP studies in the report previously cited differ on that count. The September 2008 study of 1,016 respondents between the ages of 45 and 64 indicated that 79 percent either strongly or somewhat agreed that they wanted to stay in their current homes. The December 2008 study of 104 adults between the ages of 44 and 62 showed that 48 percent would like to stay in their current homes. ProMatura Group in 2009 conducted a study of 1,241 55+ households in which respondents were asked how likely it was that they would purchase a new residence in the next 12 months. Just 4.2 percent said they were likely to buy a new home within the next 12 months. In comparison, between 2006 and 2007, 5.1 percent of households headed by someone 55 to 79 years of age moved to different residences, according to a 2007 U.S. Census Bureau study.

Another ProMatura study conducted early in 2009 polled 583 homeowners who had purchased homes in the last 2 years in an active adult community; one trend came clear from the study. For decades, active adult communities and intergenerational developments have been charging a pretty penny for upgraded lots, many of which were centered around a golf course, giving homebuyers supposedly the preferred views. In ProMatura's study, the least selected item chosen among selections for the view they wanted was a golf course. The number one preference was a view of green space, number two was a view of the mountains. However, golf course communities tend to have higher values than non-golf course communities, said Dave Schreiner, former vice president of Pulte Homes.

Another surprising trend that is surfacing, according to the ProMatura study and also an extensive study by the MetLife Mature Market Institute (MMI) and the National Association of Home Builders (NAHB), relates to active adult communities that require residents to be 55 years of age or older. The MMI/NAHB report, "Housing for the 55+ Market," cites an increase in active adult communities from 2.2 percent of all

55+ households in 2001 to 3 percent in 2007, which equates to 423,155 households. As the number of baby boomer households appropriate for active adult housing increases over the next decade, the number of residents in this type of housing is expected to increase even further. ProMatura's study supports that, predicting that demand will far outstrip supply of active adult communities through 2020, with a projected 5,000 communities available when the number of households desiring to live in one could fill 6,135 communities. One way to help fill that demand is to build age-restricted 55+ active adult enclaves within planned multigenerational communities, says John Migliaccio, director of research for MMI. See more information on active adult communities in Chap. 3.

When ProMatura's CEO, Margaret Wylde, presents results of her studies, she emphasizes over and over that her data relates to preferences for the whole United States and that her data might be the antithesis of what the market wants in each particular geographic area. Geographic and cultural differences are huge factors that builders should pay attention to, even if national studies indicate something 180 degrees different. "Not everybody wants the same thing," she said.

Dozens of studies of homeowners' preferences were conducted prior to the recession. How relevant and accurate they may be now is in question. What the authors are hearing from architects, homebuilders, economists, and other sources of information close to the homebuilding industry is that the era of conspicuous consumption of the biggest, most lavish, and ostentatious home is over. Between the hard hit that so many people have taken in their finances when they thought they had a guaranteed comfortable retirement and the changing mood of the country, homeowners are being swept up in a wave of practicality, conservatism, and saving instead of spending. "When 79 million people—nearly one third of Americans—start spending less and saving more, you know it won't be pretty," said David Welch in his *Business Week* article, "The Incredible Shrinking Boomer Economy." He continued, "According to consulting firm McKinsey, boomers' conversion to thrift could stifle the economy's hoped-for rebound and knock U. S. growth down from the 3.2 percent it has averaged since 1964 to 2.4 percent over the next 30 years," So many experts say that we're not "going back there" to "the good old days" of freewheeling spending that we all would do well to proceed accordingly. This book will help you figure out how to do that.

REFERENCES

1. Crowe, David, Chief Economist, National Association of Home Builders, presentation at Building for Boomers and Beyond Conference, Philadelphia, Pa., April 2009.
2. Groeneman, Sid, "Staying Ahead of the Curve 2007, The AARP Work and Career Study," Full Report, AARP, September 2008, Washington, D.C.
3. Helman, Ruth, Craig, Copeland, and Jack, Van Derhei, "The 2009 Retirement Confidence Survey: Economy Drives Confidence to Record Lows; Many Looking to Work Longer," EBRI Issue Brief, No. 328, April 2009.
4. "Housing for the 55+ Market: Trends and Insights on Boomers and Beyond," Report, National Association of Home Builders/MetLife Mature Market Institute, April 2009.
5. Koppen, Jean, "Effect of the Economy on Housing Choices," Research Report, AARP, Washington, D.C., February 2009.

6. Korczyk, Sophie, "Who Is Ready for Retirement, How Ready and How Can We Know?" Research Report, AARP, Washington, D.C., January 2008.

7. Nelson, Arthur C. and Robert Lang, "The Next 100 Million," *Planning,* American Planning Association, January 2007.

8. Rainville, Gerard, "AARP Bulletin: Survey on Employment Status of the 45+ Population," Executive Summary, AARP, May 2009.

9. Taylor, Paul, Rakesh Kochhar, Rich Morin, Wendy Wang, Daniel Dockterman, and Jennifer Medina, "Recession Turns a Graying Office Grayer," Pew Social & Demographic Trends Report, Pew Research Center, September 3, 2009

10. Welch, David, "The Incredible Shrinking Boomer Economy," *BusinessWeek*, Aug. 3, 2009, pp. 26–30.

11. Wylde, Margaret, *Boomers on the Horizon: Housing Preferences of the 55+ Market*, BuilderBooks (National Association of Home Builders), Washington, D.C., 2002.

12. Wylde, Margaret, "Developing Active Adult Communities: A New Model for the Baby-Boom Generation," Research study presentation at National Association of Home Builders 50+ Housing Symposium, Philadelphia, Pa. April 2009.

13. Wylde, Margaret A., *Right House, Right Place, Right Time: Community and Lifestyle Preferences of the 45+ Housing Market*, BuilderBooks (National Association of Home Builders), Washington, D.C., 2008.

14. Wylde, Margaret A., "50+ New Home Buyers: Why, Where, What and When," Research study presentation at International Builders' Show, Las Vegas, Nev, January 21, 2009.

15. Zolli, Andrew, Keynote Speech, Build Business Conference, Society for Marketing Professional Services, Las Vegas, Nev. July 18, 2009.

3 Neighborhood Types

"Don't buy the house, buy the neighborhood."

—Russian Proverb

When you are planning the types of neighborhoods in which you want to design or build, you will want to consider many factors: zoning, regional considerations, local demographics, competitive factors, and your own strengths and weaknesses.

Local zoning restrictions may actually drive you to add age-restricted housing to your repertoire. Many designers and builders who have not previously studied or worked in the business of building for aging baby boomers or seniors may suddenly find themselves having no choice if they want to go into a certain geographic area. Some municipalities, particularly in the Northeast, will only grant permits for age-restricted housing or housing that includes at least some percentage of age restrictions, because they view communities with persons over 55 as revenue-neutral or revenue-positive rather than a drain on services. Possibly more importantly, they want to limit the number of children that enter the local school system to lessen overcrowding and keep within school district budgets.

However, some of those plans backfire. Many childless boomers or empty nesters living in non–age-restricted housing want to stay close to where they currently live. So they move to a 55+ community within the same school system's grid, although that isn't a consideration for them. They often sell their homes to families with children, and that actually adds to the schools' populations.

As the authors will stress over and over, it is vital that you consider local and regional factors: regulations, such as the growth rate, traditions in architectural design, population makeup, regional customs and preferences, existing housing inventory, and competing local firms. They will try to help you understand how the coming older generation of baby boomers—77 million strong—will differ from previous older generations, but you need to include local and regional considerations or you could wind up on the wrong track.

If you find yourself thrust into the world of building for boomers and seniors for the first time, you are in for a shock. You will experience nearly the same level of shock if you are used to building for seniors only. The Internet and social networking have changed everything. Boomers are connected and they do their research, usually both before and after their first visit to your project. They will ask their pals, many of whom they have never met in person, on Facebook, Twitter, and other social networking communities, for their input. Not only will this group take 2 to 3 times as long to make a buying decision, but also they will visit your sales center 8 to 12 times and call as often with questions. Your sales staff must be prepared with the necessary information and, most important, they must be patient. Boomers will take their time as they gather and evaluate all of the information they can find.

As outlined in Chaps. 1 and 2, boomers are not now, have never been, and never will be any kind of seniors this world has seen before. Appealing to the boomers as they get older—boomers never see themselves as becoming seniors, after all—starts more importantly than ever before with the neighborhood. The old real estate saw of "Location, Location, Location!" has never been more fitting.

Keep in mind that before you can settle on designing houses for your complex, you've got to create a vision of what you'd like your community to be and for whom. Establishing a vision for your community begins with focusing on a clearly articulated segment of the huge 55+ buyer group. The temptation is to try to include everyone in the vision and ultimately satisfy no one. The same principles apply if the planning is not among your responsibilities; you have to understand the neighborhood to know what dwellings will work there.

The types of neighborhoods the authors will focus on in this chapter are (1) age-targeted, age-qualified, age-restricted; (2) mixed generations; (3) infill neighborhoods and rehabs; (4) urban downtowns, suburban city centers, transit-oriented design (TOD); (5) traditional neighborhood developments (TNDs); (6) university-related retirement communities; and (7) cohousing. Each has its own distinctive appeal to certain segments of the boomer generation, and each has its own set of guidelines and unique characteristics that you should understand.

AGE-TARGETED, AGE-RESTRICTED, AGE-QUALIFIED

You may have heard the terms age-targeted, age-restricted, and age-qualified when referring to 55+ communities. All are considered lifestyle communities, but what's the difference? In many ways, you design and build them very similarly. The legal distinctions are minimal but important.

Age-targeted communities: For any of several reasons, you may not want to limit yourself to a community totally made up of people age 55+, or you don't want to be legally bound by the restrictions associated with an official 55+ complex, but you would like to attract that age group. You can achieve that by creating an environment and offering a lifestyle desirable to persons in their fifties, sixties, or above. With the right community design, amenities, activities, and homes, prospective buyers or renters will self-select their way in. Often what appeals to older childless couples also appeals to younger, upwardly mobile childless couples, so you might find those two groups as your prospects and residents, but with similar lifestyles, they should be compatible.

Age-restricted communities: The Housing for Older Persons Act of 1995 provided for two types of age-restricted communities, amending the Fair Housing Act passed in 1968.

One type calls for at least 80 percent of the units to be occupied by—not owned by— at least one person age 55 or older. When an age 55+ owner leases a dwelling in an age-restricted community to a person younger than 55, that unit is not counted in the 80 percent required to maintain the exemption. When a 55+ neighborhood or complex is under construction, the 80 percent rule kicks in when 25 percent of the units are occupied.

The second type requires 100 percent of the units to be occupied by persons age 62 or older. In both types of age-restricted housing, the act allows for the option of a total ban of children under 18 (or any other age deemed appropriate).

State or local laws may impose additional restrictions you may have to adhere to, but at minimum you will have to comply with those in the federal law. We will be focusing on the 55+ segment.

For a property, project, or community to qualify as an age-restricted or age-qualified community and maintain that protection, it must meet the following conditions, according to the U.S. Department of Housing and Urban Development (HUD):

- Houses or dwelling units must be in one place and have some relationship to each other. They can't be scattered among non–age-restricted units.
- The complex or community owner or manager must publish and abide by a common set of rules, policies, and procedures that clearly show intent to provide housing for persons 55 or older and must be applicable to all units.
- The facility or community owner or manager must be able to verify compliance with age requirements.

Minimal bureaucracy exists for establishing 55+ communities. Federal law does not require that you submit any paperwork to qualify for the age-restricted protection but be sure to check with your state and local governments in case their rules are more stringent. And, be prepared to prove that you meet all of the requirements in case it's ever necessary.

Age-qualified communities: This is simple. The term age-qualified is the same as age-restricted. Age-restricted is more common in legal language; age-qualified is more often seen in marketing materials and is generally preferred by the industry.

Market Size

When it comes to percentages of people who move to an age-restricted community, in the past they have not been all that impressive. At best, such communities never drew more than 10 to 15 percent of the market, even when there were few alternatives for 55+ buyers. Del Webb's 2003 "Baby Boomer Study" indicated that only 7 percent of those likely to move in retirement would choose a community made up solely of people age 55 or older. The National Association of Realtors reported in 2005 that 8 percent of those considering a move in the next 5 years might consider purchasing a home in an active adult community. But even a small percentage of the massive boomer population creates a large enough market to make it worthwhile to pursue.

Lifestyle communities set the bar against which all other community types are measured. And, considering that for boomers lifestyle will be a huge consideration like for no generation before them, those percentages could significantly rise.

The Big Picture—Large Communities

Large age-restricted lifestyle communities will appeal to a national market of nearly 100 million people—counting boomers and older cohorts—and, as we said, 7 or 8 percent of them make up a pretty large, broad market.

> To develop a large lifestyle community for the 55+ crowd, a huge up-front investment of millions of dollars is needed. It is important for the designs to have a broad appeal. That means creating upward of 10 furnished model homes; 15 to 20 are better. The amenity package and clubhouse should be open and functional *before* selling the first home. All of the infrastructure, roads, parking, entry gates, and community landscaping also need to be in *before* selling the first home. This is not a model of community development for anyone but the largest developers and homebuilders (See Fig. 3-1).

Let's see what commonalities and differences exist in a handful of existing communities.

Figure 3-1

A Village of Lady Lake sales center serves the largest active adult community in the country, with 50,000 homes at build-out. (*Photo by Mike Kephart.*)

Not Just in the Sunbelt—Sun City Huntley

Originally, most huge communities were in the Sunbelt, but they now can be found all over the United States. Del Webb, long the leader in large lifestyle communities, developed their first non–Sunbelt Sun City in Huntley, Illinois. Sales of new active adult homes in the Chicago area grew 49 percent from 2004 to 2005, according to Tracy Cross of Tracy Cross & Associates, Schaumberg, Illinois. And that was before the baby boomers started turning 60!

When Sun City Huntley opened in 1998, the only competitive communities in the area were Cambridge Homes' Carillon developments. Now there are upward of 30 more active adult projects in the area.

Sun City Huntley was planned for a total of 5,800 homes on 2,150 acres. Amenities include a 27-hole golf course, a 94,000-square-foot recreation center, tennis courts, indoor and outdoor swimming pools, and a full complement of parks and walking and hiking trails. To attract large numbers of age 55+ homebuyers, these types of communities are designed to have something for as broad a niche of buyers as possible—in other words, something for everyone.

The community plan is a conventional curvilinear street system, locating the front-loaded houses along the golf course fairways. This design of the site oriented the rear yards toward golf courses or open spaces, so great efforts were made to locate family rooms, nooks, and master bedrooms on the rear to take advantage of the views. With many different home plans, each was designed to reach a certain segment of the market.

Most people, for instance, preferred the master bedroom to be located away from the other bedrooms, guest room, study, etc. Other buyers liked having at least one additional bedroom close to the master bedroom for the snoring mate to retreat to when necessary, or for other reasons.

A New Twist—Solivita

Home designs in larger communities are typically suburban in style, with attached garages on the fronts of homes and large yards in the rear. Houses are usually wider than they are deep to create as large a presence as possible facing the street. The plans are designed to focus on the views of open space, golf courses, or lakes when looking out from the rear of the homes. In fact, the view catches the eye as soon as you enter the most popular models. These "see-through" houses leave little to discover after the visitor recovers from the impact of seeing the entire home and surrounding environment all in one glance (see Fig. 3-2). This design concept is strongly preferred, but when offering so many models, variety is necessary. Since most people choose the see-through models, the majority of homes are designed that way, but some people like their nook up front, so as to be able to see approaching visitors. Yet others prefer a more formal plan arrangement with clearer room definition. See Chap. 6 for more on floor plans.

> In the course of doing research for this book, the authors found several communities that had put their own stamp on the "Sun City" concept and come up with some new twists. A brief highlight on a few of these communities should give you an idea of how many ways the fundamental concept shows up in the market.

Figure 3-2

A "see-through" floor plan presents views of the rear open space upon entering, thereby capturing outdoor space as part of the home. (*Drawing courtesy of KEPHART.*)

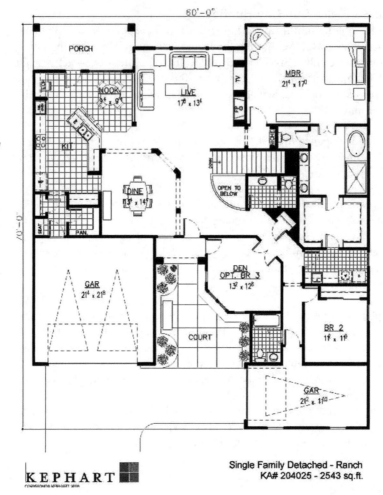

KEPHART

Single Family Detached - Ranch
KA# 204025 - 2543 sq.ft.

Solivita, an Avatar community in Poinciana, Florida, is a lifestyle community seemingly patterned in the Sun City mold until you look at the community amenities. The fitness center, the arts center and studios, and all of the other amenities are strung along both sides of a commercial street. Restaurants, offices, stores, and service businesses, such as travel agents and golf car outlets are woven into the busy street scene, creating the colorful impression of a downtown in a small Caribbean city (Fig. 3-3).

> Solivita's Main Street style of amenities, versus the usual clubhouse after clubhouse so common in active adult communities, just seems to present more of a "normal" atmosphere for day-to-day living rather than having a few services squished into a clubhouse or having residents go outside the active adult community for everything they need.

The homes at Solivita are similar to those found at any lifestyle community in their design concept. Views out to golf courses or open spaces fill the eye when entering each of the homes, and plans including great rooms outnumber plans with both living rooms and family rooms. There are 16 models of single-family, one-story homes offered at

Figure 3-3
The clubhouse at Solivita isn't one structure but instead a group of buildings along a street. (*Photo by Mike Kephart.*)

Solivita, with four more in their courtyard paired homes. Uncharacteristically in lifestyle communities, these homes are distinctly narrow and deep in shape to minimize the street length devoted to each home, perhaps a concession to the rising land prices that encourage higher density solutions. These courtyard homes are not small; they range from 1,972 to 2,699 square feet, while the courtyards are compact. Two models even have additional bedrooms on a second level. The courtyards offer only one garage space plus a short "golf car" space, recognizing the planning that provides for golf cars throughout the community.

The Mother of Them All—The Villages

The Villages in Lady Lake, Florida, (see Fig. 3-1) approximately 50 miles north of Orlando, dwarfs nearly every other large active adult community. The Villages development spans five counties over untold acres of land in central Florida. About 30,000 people currently live and play in The Villages, and it's expected that the total population will exceed 100,000 at build-out. There are 20 nine-hole golf courses, free to all residents and eight country-club courses that are open to all on a daily-fee basis. The feature course, designed by Arnold Palmer, costs only $50 per 18-hole round.

Two town squares, each fully developed to offer a large range of restaurants, retail, services, and entertainment, are available both to residents and those outside the community. One is Mediterranean in style; the other has a nautical theme, taking a design cue from the lakes adjoining the center. A major shopping center is located just outside the community boundaries in order to draw businesses that would not normally venture into a closed community.

Thirty gated "villages" are planned. To get around, residents can literally travel anywhere within The Villages on their customized golf cars. Each village contains its own amenities, which are also open to anyone living in any of The Villages. There is a polo field, stables, and a viewing pavilion for watching matches. Swimming centers are scattered about the community and each village also has its own pool.

> When author Kephart visited The Villages, the softball complex of four diamonds was filled with players and fans. And, pickle ball is one of the most popular sports there, allowing anyone, even the nonathlete, to play and have fun. It's a tennis-like game, played with a lighter ball on a smaller court, but the countless courts available at The Villages were all alive with laughing, sweating players.

Homes in The Villages are unremarkable. They follow the tradition of other lifestyle communities and are designed with front-loaded garages and views out the back to open space, golf, or water.

> Author Kephart opines: I realize that the community concept is "all about lifestyle," but the architecture should at least equal the quality of that lifestyle. In my opinion, the home designs in The Villages fall short of that goal.

Since The Villages community is being developed and built by a family company, there is no opportunity for small builders to participate, but it's different in many communities around the country.

Mega Communities Aren't for Everyone

One size doesn't fit all, nor is it always wise to try to offer something for everyone. Segmenting homes to fit varying lifestyles is going to be more important than ever as always-independent, individualistic boomers reach their mid-fifties and above. Certainly not everyone wants to live in mega communities. Even some of the large homebuilders are thinking smaller when it comes to their newer communities designed for age 55+. That includes Pulte Homes and its active-adult-oriented Del Webb subsidary. For example, Del Webb's Corte Bella, a gated 55+ country club community adjacent to Sun City West in Arizona, has only 165 homes ranging from 1,155 to 3,027 square feet.

Smaller-Scale—Arbors at Bridgewater Crossing

You don't necessarily have to be a large homebuilder to create a lifestyle community. Parkview Homes of Strongsville, Ohio, built and sold homes in its 200-home project called the Arbors at Bridgewater Crossing in Brunswick Hills, Ohio, from 2006 to 2008 (see Fig. 3-4). The development offers single-family homes plus attached or paired homes, a 4,000 square feet community activity center, and a site plan that boasts of walking trails, lakes, and neighborhood and community parks throughout. The homes in the Arbors are modestly priced and sized in response to the boomer concerns about affordability.

Though the development is not age-restricted, it is targeted to appeal to active adults.

> "Age-targeted" is not a term that you can use in any public sense or in marketing messages, as it can be construed as discriminatory. But designing projects with the features that appeal to boomers is perfectly acceptable and will naturally attract the residents you have in mind.

Figure 3-4
The Arbors at Bridgewater Crossing project in Ohio boasts of 200 homes. (*Courtesy of KEPHART.*)

In this case, the builder made every effort to provide something for all residents just as the large-scale communities do, but in a way that would be more attractive to boomers than to families. There's a full-time activities director/fitness coach who conducts pool exercise and gym/aerobics classes and also arranges outings and tours to local sights and athletic and cultural events. Thirty people were getting ready to board a bus to a Cleveland Indians game the last time author Kephart stopped by.

Parkview's Richard Puzzitiello, Jr., said, "There are things we would do differently, of course. We have too much attached product and the cluster sites have not met sales expectations, but overall the concept worked. Committing up front to build the clubhouse and all of the furnished models was a great decision. Oversizing the clubhouse, pool and fitness center and including card rooms and multipurpose rooms gave our activities director the tools to make this lifestyle work. We would tweak the details and do it again if we could find the right location."

Communities within Communities
Some larger communities have distinctive, smaller ones within. The 3,500-acre Talega community in San Clemente, California, includes within the master plan a small, age-restricted, active adult neighborhood of about 200 homes, called Talega Gallery, a project of Standard Pacific Homes. It is a gated enclave with its own community association and its own club, including a pool and year-round heated hot tub, fitness center, nine-hole putting green, library, and ballroom. People under age 18 are not permitted in the Talega Gallery pool and clubhouse, but the two nearby Talega community pools and recreation facilities are quite child friendly for residents with visiting grandchildren.

Del Webb's Solera at Johnson Ranch in the Queen Creek area, southeast of Phoenix, has 726 single-story homes from 1,398 to 1,905 square feet and is billed as "a collection of neighborhoods." It has its own 10,000-square-foot recreation center and outdoor pool and spa, but residents also have access to the pools and other amenities in the larger Johnson Ranch development.

Pulte Homes' 2,648-acre Anthem community in Broomfield, Colorado, north of Denver, is split up into three nearly equal parts: the age-restricted Anthem Ranch by Del Webb, Anthem Highlands for all ages, and a planned mixed-use office, retail, commercial, and town center.

Active Adult–Oriented but Open to All—Within Ave Maria and Anthem Parkside
The 11,000 homes planned within the 5,000 acres comprising the newly created town of Ave Maria, Florida, near Naples, are divided into six distinct communities, including one clearly aimed at active adults, even though it is open to all ages (see Fig. 3-5). Ave Maria is a large development that may be the first of its size in the United States to be centered on a spiritual base. Life in Ave Maria revolves around a new Roman Catholic university called Ave Maria University and a 100-foot-tall, $24 million church named the Ave Maria Oratory, named oratory because until March 2008 it was not affiliated with the church and could not serve some of the functions of the Roman Catholic Church. Founder of the university and cofounder of the town is Tom Monaghan, who also started Domino's Pizza and headed it until 1998.

The whole development is open to all ages. But the northern part of the town has schools, sports fields, pools, and other active amenities. The southern part of the development has more of a sedate, sophisticated atmosphere, with single-story homes built around a golf course by Del Webb. It's not age-restricted, but the people who are looking to buy there are going to choose homes in the area that has the amenities they want.

In the other U.S. Sunbelt—in the Southwest—Del Webb now has niche neighborhoods within their larger communities to target specific segments of buyers. Its Anthem Parkside community in Arizona, which has no restrictions on age throughout its planned 14,000 homes, is building schools for resident children in kindergarten through high school, but it also includes a section clearly designed for an older market without children. To reinforce the dual focus, early marketing materials described Anthem Parkside as "part kid kingdom and part grown-up getaway."

The area within Anthem Parkside that would appeal to the 55+ market is age-targeted, not age-restricted. It makes up about one-third of the community. Homes and amenities are distinctly different from the family section in ways that clearly divide the two.

Homes within the family-oriented section include both one-story and two-story types. Bedrooms are clumped together so parents can hear and keep an eye on the kids nearby. Amenities are family-friendly: a three-story rock-climbing wall and a wave pool that obviously draws kids; and a full-size indoor basketball court, along with the ubiquitous fitness center. In the area clearly aimed toward the 55+ crowd, single-story homes abound. The master bedroom is on one side and the other bedrooms are on the other side of the house, which appeals to empty nesters and couples or singles with grown children or

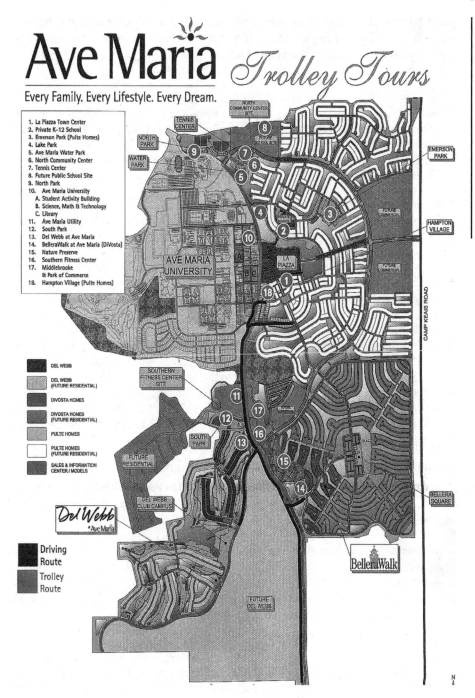

Ave Maria
Trolley Tours

Every Family. Every Lifestyle. Every Dream.

1. La Piazza Town Center
2. Private K–12 School
3. Emerson Park (Pulte Homes)
4. Lake Park
5. Ave Maria Water Park
6. North Community Center
7. Tennis Center
8. Future Public School Site
9. North Park
10. Ave Maria University
    A. Student Activity Building
    B. Science, Math & Technology
    C. Library
11. Ave Maria Utility
12. South Park
13. Del Webb at Ave Maria
14. BelleraWalk at Ave Maria (DiVosta)
15. Nature Preserve
16. Southern Fitness Center
17. Middlebrooke
    & Park of Commerce
18. Hampton Village (Pulte Homes)

DEL WEBB
DEL WEBB (FUTURE RESIDENTIAL)
DIVOSTA HOMES
DIVOSTA HOMES (FUTURE RESIDENTIAL)
PULTE HOMES
PULTE HOMES (FUTURE RESIDENTIAL)
SALES & INFORMATION CENTER / MODELS

Del Webb
at Ave Maria

Driving Route
Trolley Route

Figure 3-5
Ave Maria Florida's new town plan includes a university, a town center, and several residential neighborhoods, each defined by the amenities within that neighborhood. (*Map courtesy of developer Barron Collier, Brian Gable.*)

grandchildren. The amenities are similar to those in the family-oriented section but more sedate. The leisure pool has a walk-in entry that accommodates people with disabilities and has no steps or ladders. The active adult area is not gated but has a wall that wraps around it and forms an entryway of sorts on the northwest end of the development.

> The active adult area has its own pool and clubhouse, which anybody can use, but for families, it's like, who's going to go over there and use that? The area has its own implied privacy. The authors' advice for anyone who wants to create an age-targeted community within a larger multigenerational community is to clearly emphasize the design features and amenities that older buyers will value and appreciate. Then people can make their own choices.

Residents of the enclaves may share use of the master community amenities plus enjoy additional amenities in their own neighborhood. The size of the enclave of homes that appeal to the 55+ market may be as small as a city block or as large as half of the development.

Choices Make the Difference   If you choose not to age-qualify these enclaves, a naturally occurring mix of buyers will develop. Residents can mix as they like if the amenities are designed with the 55+ person in mind. Possibly most important of all is to offer *choices*.

Single-story homes or two-story homes with elevators plus maintenance-free living will attract older boomers and fewer families. Most of the non–55+ homebuyers in a neighborhood designed to attract 55+ buyers are typically young professionals or singles who travel for business. Families are not drawn to the more costly single-story homes or to the small yards with little room for young children and pets. Members of the 55+ crowd are willing to pay extra for a one-story, single-family detached home that is fully maintained for them—regardless of yard size. See Chaps. 6 through 9 for more on how to design and build homes that will appeal to boomers as they pass age 50.

Ideally residents would be able to choose from among three kinds of walking experiences: slow walks with friends, strenuous walks for exercise, and quiet walks for meditation. A series of connecting loops will allow a person to choose the distance and exertion level they want for any particular day.

A separate wellness/fitness facility for the 55+ crowd will give them special activities like pool exercise classes that many older residents would enjoy but younger people wouldn't. An added benefit is that having their own fitness facility gives boomers some peace and quiet away from a noisy family pool or gym. Boomers may also enjoy cards with friends, hobbies, or games not so popular with younger people.

> In the 55+, age-restricted community where author Schriener lives, when Wii became popular, her complex had one in the club room. Residents enjoyed it immensely and were as competitive as anyone. But if they had had to compete with younger folks or worry about being ridiculed by them, they would have probably stayed away and missed out on a wonderful activity that kept them current and, just as importantly, kept them moving.

For mixing with all generations, common gathering places within a larger community are coffee shops, restaurants, post office, parks, convenience stores, and community activities and performances.

Intergenerational communities use descriptions of their communities and accompanying lifestyle choices in place of old words like "seniors," "active adults," or "retirement living." See more about marketing to boomers in Chap. 12.

MIXED GENERATIONS

Communities that include all age groups are appealing to many 50+ buyers. Some 50+ buyers don't like being segregated with "old folks" or even people their own age. Some just like the diversity and normalcy of living among people of all ages. Some want to keep their families close together. The cultural heritages of many immigrant groups, even if they are in their second or third "American generation," include a strong connection to family. Some boomers living among multiple generations will want to be very near to their children and grandchildren but no longer want a family-sized home.

There are two fundamental ways to include all ages within a community plan and still provide the features and amenities that the 55+ homeowners particularly enjoy, constituting a modified intergenerational layout. One way is to develop an age-restricted or age-targeted enclave within a master-planned community. This was described in the previous section. But creating enclaves isn't the only way to approach the challenge of achieving a balance between giving boomers the lifestyle they desire and yet not segregating them. This can also be accomplished by optimizing the locations of the homes within the larger community, by sprinkling age-targeted homes in with family-oriented homes.

One way of giving boomers what they want and need and yet keep them integrated within an intergenerational community is to mix larger and smaller homes very finely throughout a development. An example is the "quilting" lot pattern of Denver's signature Stapleton redevelopment on the huge site of the former Stapleton International Airport, which closed and was replaced by Denver International Airport in 1994. Stapleton, on 4,700 acres, bills itself on its Web site as "the largest urban redevelopment community in the nation."

The intent of the original plan at Stapleton was to mix home sizes and price ranges throughout the community in the way some older neighborhoods evolve over time. Small lots were originally platted next door to bigger lots, and corners were reserved for manor homes or courtyard clusters. The concept was later modified to group same-sized lots block by block. Homebuilders were having trouble staging their work and marketing their homes when construction was spread out over several city blocks with two, three, or even just one home on each block.

The "quilting" plan differs from traditional plans where higher-priced homes are separated from the lower-priced ones, requiring buyers to make a social and political decision as well as an economic one. Quilting can be very fine, as in house-by-house, or very coarse, as in the old "pod" system of planning, and everything in between. There is no precise definition, only a change in scale of the mix and the complete elimination of giant neighborhoods of the same lot sizes, house types, and designs.

INFILL NEIGHBORHOODS AND REHABS

The remodeling of existing homes is the most common method of updating neighborhoods, but sometimes, larger changes outside the control of homeowners, such as the closing of a neighborhood school, take place. City plans also change, which can impact neighborhoods.

Of course when homeowners remodel old homes into the new homes they want, that may be the best choice for many of them who do want to stay in their existing homes and neighborhoods. In author Kephart's neighborhood inside the Denver city limits, for example, city zoning permits the addition of a second home on each lot. This second home can be sold or rented. This is not a choice everywhere, but for older residents living in neighborhoods that are becoming too valuable to stay in, it might be an option.

However, remodeling often just won't do the job or is too costly. For some, an infill location for a new house is the answer. Builders want to build homes that people will buy; new homes with modern kitchens and baths and the latest energy-conserving features are highly valued in old neighborhoods. Opportunities to build in an infill location arise when an old school closes or a business moves out of a neighborhood. Sometimes the opportunities are created by city initiatives such as a light rail expansion with a stop nearby, or when a city develops new neighborhood plans.

Opportunities

One of the Denver-based KEPHART firm's recent infill projects was the result of both a new light rail stop and a private school closing. Negotiations between the builder/developer and the neighborhood residents created a plan that included renovation of one of the historically significant buildings into condominiums plus the addition of some single-family homes and townhouses. Neighborhood character and comparable home sizes were integral in the project design. Every effort was made to satisfy current and new neighborhood residents' desires. The classy project fit perfectly into the neighborhood. The beautiful old brick building, former home of the Denver Academy prep school, was refurbished and converted to mid-priced condominiums, which fully sold out.

A new home in the old familiar neighborhood could be considered a nearly perfect solution for the overwhelming majority of 50+ homebuyers. Old friends are there, as is the support that they and other neighbors provide. The new homes can be up-to-date and more appropriately designed for a new lifestyle. One-story homes would complete the ideal, but here is where an obstacle looms—land cost. Higher densities are often needed to bring that cost into line.

Single-story homes in tight clusters may be able to achieve the density goal, but often do not. Personal elevators in two- or three-level homes can aid the creation of home designs that are as easy to live in as the one-story ones, and they use one-third to half of the land area. Multifamily housing is an option if the home sizes and quality equal or exceed that of single-family houses, or if they are simply more affordable, a common role played by multifamily development. Multifamily proposals tend to face a higher degree of resistance from neighborhoods, however.

In California, where land costs are as high as anywhere, architects have devised creative ways to replace one or two existing homes that have outlived their usefulness with three or more new ones.

Larger infill parcels of land can sometimes be found in areas of change. In the case of One Cherry Lane in the Denver Technological Center, the original office park plan was revised after most of the development had filled up with office buildings and commercial campuses. The plan revision opened up areas for residential development and Esprit Homes was able to secure 25 acres that the developer rezoned for a luxury community of 86 single-family homes.

One-story homes sized from 2,600 to 3,500 square feet were designed for affluent 50+ home buyers who also owned at least one other home and were currently members of one or more country clubs where they played golf and socialized.

Four model homes were designed, each with a stone or brick exterior. Interior courtyards have zero-step entries to the homes, achieved by raising the courtyard floor up to the same level as the interior. This allows the courtyards to be designed as outdoor rooms. Living spaces orient toward these rooms which provide very private outdoor living opportunities.

These homes were repeated, with custom touches, throughout the community. Homes were priced from $800,000 to $1.3 million in 2007. The design program was developed largely from findings recorded from a series of buyer focus groups.

It was a very compact, high-density neighborhood of patio homes with nearly nonexistent backyards. They are located in the middle of an active tech center of multistory office buildings, restaurants, and shops, which were attractive to prospective homebuyers because they were so close by, even if they couldn't walk to them.

Quail Hollow in Erie, Ohio, is another example of opportunity presented with a change to an existing master plan. An island of land surrounded by the golf course was originally intended for large custom homes, but no builder had elected to buy the property and carry out the plan. Bill Martin, president of Bainbridge Homes, negotiated a purchase price based on being able to double or triple the density and build cluster homes for the 50+ market. Six- to eight-home clusters made the density of five homes per acre feel far less intense. Trees were preserved around the perimeter of the property between clusters and in guest parking islands, retaining the forested character of the site. Floor plans are narrow at the street front, and step back to their widest at the rear of the homes, which allowed the clusters of homes to be pulled close together at the cluster drive and to widen out to open space in the rear.

> A simple way to double the density and cut land cost in half is to build two houses on one lot. Depending on the lot sizes in your region, it may be necessary to also borrow from the two-story active adult concept and build in elevators.

The front carriage homes with standard residential elevators are above garage spaces for both houses. Guest parking is provided in 90 degree parking nodes clustered around a central pedestrian walkway leading to all four houses on two paired lots.

It is also age-inclusive, which is preferred by a majority of 50+ homebuyers. The one-story homes at the back of the lots are perfect for those age 50+, while the carriage homes above garages are intended for young professionals or couples with or without young children. Elevators, of course, are not necessary to appeal to these younger buyers, but here is one place an elevator option may make sense.

The trade-offs in order to gain density and lower costs with this idea are: one home is above shared garages; the back home has little street connection visually; and its garage is detached. Benefits in lifestyle include the single-family detached character and the quiet seclusion of the homes in the rear, all at a significantly reduced price as compared to the typical single-family homes on their own lots with large yards to care for.

Infill can be done at any scale from one house on one lot to large developments on land reclaimed by either a political process, such as closing a military base; or a physical process, such as filling in bays, lakes, or ravines, or cleaning up a polluted site. The great secret to success with infill development is in finding a way to use a piece of land that no one else has managed to uncover.

Developing an infill community with the right home designs in a strong market area will usually assure success. Competition is minimal or nonexistent, at least for the duration of the sales period for your community. You are able to realize higher prices, within limits, from your completed homes. This all comes at its own costs, however. The following example was costly, due to having to correct the problems with the property, and becoming time-consuming, taking years to accomplish. Many sites share both of these obstacles, making the larger parcels far more attractive if investors are willing to provide financing and are willing to wait years for a return on their money.

Challenges Ahead

One of author Kephart's projects, the Courtyards at Rolling Hills in metropolitan Denver, is a classic example of the problems presented by an unfavorable political environment. Throw in some major engineering issues and it's easier to understand why this prime view site in the exclusive Rolling Hills Country Club neighborhood remained undeveloped.

The property is a long, narrow site parallel to a moderately busy street. The grade drops 25 feet from the street in the 200-foot depth of the site, offering walk-out basement opportunities but creating steep access road grades. The county would permit only one access point to the adjacent street and, over a year into the review process, decided that any on-site intersection must be over half of the depth of the site away from the access point. This reversal in the process sent everyone back to the drawing board.

Ten single-family, one-story patio homes, at a density of eight homes per acre, were the final solution to a long and contentious political process, a common occurrence with infill in quality neighborhoods.

Colorado Land and Building Company, the builder/developer, wisely elected to make the surrounding neighbors part of the process from early on. Because of the neighbors' involvement and the sway they held over the county commissioners' final decision, they were the project's best spokespeople through the rest of the process. Townhouses or condominiums were eliminated from consideration early in the process, much to the neighbors' delight, and smaller, more affordable homes were also rejected as a possibility for the same reasons.

Finally, ten single-family, one-story patio homes for age 50+ homebuyers was the chosen scheme. The neighbors and the commissioners supported the proposal, but the county engineers weren't satisfied. In the end, the builder contributed to the cost of a new off-site sewer line extension and provided space for an easement through the property to serve land areas to the north in the future. All told, this process took over three years and cost many thousands of extra dollars

Multistories

Multistory homes in infill locations are more affordable and with good design can fit comfortably into the life of a 50+ couple. The two-story active adult home concept is perfect for infill sites. Two-story homes use less land than the same-sized one-story home and, with good design, a two-story home can be as convenient to live in as a single-story ranch. An integral elevator eliminates any concern about access later in life and is convenient for moving furniture, luggage and heavy items even when mobility isn't a concern for the family members.

The idea can be stretched upward into three-story designs for higher-density locations, as it often is in townhouses in urban locations. A garage and additional storage and utility space at the lower level is easy and convenient to use and, with an elevator, the ground coverage is reduced to its minimum.

URBAN DOWNTOWNS, SUBURBAN CITY CENTERS, TRADITIONAL NEIGHBORHOOD DEVELOPMENTS, AND TRANSIT-ORIENTED DESIGN

Baby boomers more than any previous generation want to be close to the action, near all of the conveniences, and surrounded by people. No being put out to pasture for them! Previous 55+ communities often were situated way outside the city limits, often virtually isolating residents. Boomers will not put up with that. They may have to sacrifice views of golf courses and green expanses, but they will gladly trade them for the ability to get to a cultural event in just a few minutes or walk to restaurants, stores, and doctors' appointments. Few boomers dream of whiling away their hours in a rocking chair watching the grass grow anyway, even if it's a lovely view.

What urban downtowns, suburban city centers, traditional neighborhood developments (TNDs) and transit-oriented design (TOD) have in common is fulfillment of those boomers' desires to live in these activity zones, which are full of conveniences and culture, rather than on the fringes.

Urban Downtowns

Significant numbers of 55+ homebuyers are moving downtown. For some of them, it's a return to city living; for others, it's a move in from the suburbs. They are choosing contemporary lofts and more conventional condos, but the common draw is the action, culture, sports, entertainment, restaurants, and the ability to walk nearly everywhere they want or need to go in a single day.

A 2005 Del Webb study of baby boomers indicated that 30 percent of those considering purchasing a home in an age-qualified active adult community prefer to be in an urban locale. The National Association of Realtors put out a report saying that 12 percent of 55+ boomers prefer an urban center as a place to retire. The Del Webb report went on to say that family incomes play a strong role in where people choose to live. Boomers with annual household incomes over $100,000 are more likely to choose an urban area, the cost of living being a likely factor. Urban living will not replace the Sunbelt as an attraction, but as Dave Schreiner, former vice president of Del Webb, said, "Small niches are now big trends."

Data from the 2000 census, the most recent one from the U.S. Census Bureau, shows that between 1990 and 2000, of the 24 cities analyzed, six showed population growth in their downtown while showing a population decline in their city. Twelve cities showed growth both in their downtown and their city. Only four were down in population in their downtown and up in their city. And two showed declines in both their downtown and their city.

A study of 24 large American cities (not exactly the same ones as in the census study) by the Brookings Institution and Fannie Mae found that a reversal of the trend of people moving out of downtowns was under way. Eighteen of the 24 cities showed growth. The study concluded that the downtown population of Chicago was expected to increase by one-third between 1998 and 2010. The downtown population in Houston would increase by 3.5 times, more than double in Seattle, more than triple in Cleveland, and nearly triple in Denver.

Boomers comprise 32.8 percent of the population, according to the Downtown Denver Partnership. Other highly ranked boomer cities are Seattle with boomers comprising 31.5 percent of its population, Atlanta with 31.4 percent, Washington, D.C., with 31.4 percent, Portland, Oregon, with 31.4 percent and San Francisco with 30.8 percent. Inside Denver's urban core, nearly 14,000 living units were built between 1990 and 2006. By some estimates, over 50 percent of those buyers are 55+ singles or couples. That figure is an estimate because builders and developers didn't track age groups except to compute average ages in downtown. With one-third of the entire city population being boomers, an estimate that 50 percent of the downtown population is made up of boomers may actually be conservative. The 2000 census reported a total of 13 percent of families in Denver with one or more children, and fewer than 5 percent of families living at the central core of the business with any children.

The Brookings report indicated a population in Denver's downtown of 9,000, according to the 2000 census. That was technically correct at the time, but the area that most Denver residents think of as downtown includes: "LoDo," the Central Platte Valley, Ballpark,

Five Points, the Golden Triangle, and other center city neighborhoods. That area contained 83,000 residents, per the 2000 census.

Other cities no doubt have similar clarifications to make. It's important to understand these details before considering building in a downtown area in any city. To put this "mini-trend" of downtown growth in perspective, however, the growth rate in downtown Denver was minuscule when compared to suburban growth during the same period.

When building in an urban center with limited space, you may be tempted to focus on the exteriors, common spaces, or amenities. Don't make the mistake of giving short shrift to floor plans. The plain vanilla floor plans that result when trying to design for everyone don't really excite anyone. But you don't have to go too far afield to create different types of downtown spaces for different segments of the population.

It's not much of a surprise that 55+ buyers were drawn to the Artisan Lofts. These types of buildings can offer specific designs and features for each buyer segment and create a better product for everyone. "No walls between you and your imagination" was the marketing slogan for the project. Offering wide-open spaces a couple could finish as they desire should have great appeal to the hands-on consumer that wants to be involved. This strategy was a great success at the Artisan Lofts and is one to consider when designing any home for the 55+ market.

A common fitness center, indoor pool, multipurpose rooms, and club room can all be added at lobby level or on the penthouse level to enrich the lifestyle. Individual storage rooms at the garage level will answer one common objection to condominium living: inadequate storage. From this basic menu of features and services, the design vision can look further for a concept that transcends the physical elements to create a design that appeals to people on an emotional level.

A more affordable vision could incorporate a lifecenter (fitness and health) concept as the defining feature. This center, open to the public for memberships, could serve the entire downtown market. Homes could be built above and around the lifecenter since the fitness areas require few windows and can be constructed on several levels. Remember that baby boomers will never get old and they see themselves as vital, active people, so this particular concept may be coming into its own as a viable approach.

David Hall, founder of LifeCenter Plus in Hudson, Ohio, has created a model fitness and health-delivery system he believes can be replicated in 50+ communities. LifeCenter Plus in Hudson has 187 staff, 6,500 full members, and serves more than 1,200 people per day. It's a breeder facility that will offer training and consultation to builders and developers of 50+ communities. The following is his description of the concept for improving the life and health of age 50+ people:

"LifeCenter Plus is a 103,000-square-foot unique health and fitness center ranked in the top 100 in North America and located in Hudson, Ohio. LifeCenter Plus is a family-focused center with an older adult emphasis. A diverse array of programs is offered in 180 classes each week by a staff of 187 people, serving an average of 1,200 members per day, with one-third of those served being 70+ years old." Hall is also the owner of a real estate brokerage and development company with 30+ years of experience in site selection. He also adds postgraduate studies in gerontology to aid him in working with 55+ market adaptations.

See the Appendix to see Hall's full description of his facilities and their mission for the 50+ boomers, which he wrote especially for this book.

Nearly every type of age-targeted or age-qualified community described in this book has the potential to be the seed for a vision for a downtown building. A downtown housing development can even take its inspiration from a small town, as Los Angeles Eco-Village has done. Residents have planted trees on the road, drawn huge murals in the street to calm traffic, built a food co-op, and they even hold workshops and eat group meals in the middle of the street. Downtown communities can be anything: exclusive; a cohousing community; alternative; or spiritual. Downtown homes can be "reverse second homes" for those who have moved to the Sunbelt but want a place back home to use when visiting family and friends.

Suburban City Centers

Suburban city centers are much like downtowns, only with parking and, usually, cleaner air. Many suburban communities started out as bedroom satellites to central metropolitan cities and grew to become city centers in their own right. With more room to work with than downtowns, suburban city centers offer a wider variety of stores, restaurants, food markets, movie theaters, fitness centers, yoga studios, and coffee shops that serve as community living rooms. Services abound, including dry cleaning, shoe repair, driver's license renewal, utility service centers, etc.

Jim Chapman, who heads Jim Chapman Communities Inc., Atlanta, has a business model that's based on creating infill communities: 20 to 50 acres in a very highly populated section of metro Atlanta, outside Atlanta's main beltway. He loves to follow a developer who couldn't get the approval to build apartments or townhouses. "The municipalities love to give you that zoning [for age 55+] because they know they're not going to have anybody from there going to their schools," he said. He likes to build in densely populated areas so his developments draw enough people in a 10- to 15-mile radius who want to stay in the area rather than go miles away to a more isolated active adult community. The former custom homebuilder added, "You want them to be in a special infill spot that has wonderful things abutting and adjacent to it but nothing 'bad,' so most of our communities are surrounded by nice subdivisions or benign commercial uses. It's a nice, quiet, pastoral setting, yet also in the thick of things."

Housing is more diverse in suburban city centers than in downtowns, offering boomers who are looking for more options of maintenance-free homes. Many city centers now include a mix of housing options: lofts above stores accessed by elevators, freestanding townhouses, condominiums, and maybe some single-family homes at the fringes of the center.

Today, Schaumburg, which has a population of about 75,000 in a 19-square-mile area, has a town square with a major food store and lots of retail choices, all within walking distances of each other. There are government offices, theaters, medical services, an ice rink, and a wide choice of restaurants and specialty shops. See more about town centers in the "Traditional Neighborhood Developments" section.

The missing ingredient in most suburban centers is housing. Though there are lots of housing choices close by, none are located in the town square proper.

From Villa Italia to Belmar

In 1999, Continuum Partners bought the Vila Italia mall and worked with the City of Lakewood to create a fresh new development concept that finely mixed a variety of uses into an extension of the town center across the street. See more on town centers in the section "Traditional Neighborhood Developments."

The town center itself was a product of a couple of decades of planning and execution to modernize and revitalize the area. Two factors worked strongly in their favor. They had bought a large piece of land uninterrupted by streets or roads, and they had assembled all of their government functions there in various contemporary buildings. The lowlands were developed into a wetlands restoration area and heritage center. The properties at the edges of the two bordering arterial streets were reserved for commercial development and multifamily housing. Today the public buildings, surrounded by public plazas and open spaces, are the focus of the center, and the transformation of Villa Italia into Belmar certainly enhanced the area even further.

In 2001, Villa Italia was mostly demolished to create the space for Belmar. Today there are offices above retail, lofts above stores, townhouses, condos, apartments, a wide variety of name-brand retail outlets, grocery stores, and restaurants galore, which are packed every evening. A cineplex is the centerpiece in the heart of the center, across the street from an ice skating rink. The Belmar/Lakewood City Center area is the place to be in Lakewood, something no residential community alone can create.

A well-designed and planned suburban city center like Belmar offers a complete lifestyle with a mix of housing and prices. While there are no age-restricted neighborhoods within this development, a significant proportion of homebuyers are over 50 years of age. Many of those older homebuyers will choose Belmar because it's close to their old neighborhood. Many others will come who may have considered downtown, which is also rich in housing choices appealing to the older buyer. Housing in the center is made up of townhouses, lofts, condominiums, or other multifamily configurations, many of which are made more accessible by the inclusion of elevators. These compact building forms use less land than houses and are more compatible in scale with the nearby commercial buildings.

Traditional Neighborhood Developments

Boomers have seen a lot of changes in neighborhoods in the years since they were carefree kids. They may be nostalgic for the types of neighborhoods in which they grew up, but they are also interested in finding the best all-around environment for everyday life. Traditional neighborhood developments (TNDs) offer an appealing blend of both.

Traditional neighborhood development planning is called *new urbanism*. Planners look for elements of old towns that, when blended with modern-day elements, will create optimal communities. TND planning was born out of a desire to replace sprawling cookie-cutter suburbs with neighborhoods that have the advantages of being close-in to reduce travel times, along with the feeling of community fostered by plentiful opportunities for casual interaction among residents. These mixed-use neighborhoods generally are compact, dense, and close-knit. People want to live there.

TNDs are relatively new, despite having their roots in the past. In 1993, 170 architects and planners got together to address the issue of creating livable, desirable communities while preserving the environment. They formed the Congress for the New Urbanism (CNU), which has taken off like a rocket. CNU now has more than 3,000 members in nearly all states and 20 countries. CNU cites more than 200 TNDs either being built or already built in the United States. TNDs for the 55+ homebuyer as an intentional market, however, are newly emerging. The compact nature of a TND has forced most homes within existing TND neighborhoods into two or three stories. Older TNDs also are loaded with stairs up to porches in order to raise homes above street level for privacy when the walk is so close by. Other than the stairs and multilevel homes, a TND should make a very comfortable community model for those boomers in their fifties and sixties.

The design elements that make up a TND are

- A variety of housing types, ranging from single-family to apartments, in all price ranges
- A mix of residences and commercial, academic, and civic buildings
- A balance of public and private spaces, with plentiful parks and gathering places
- Interconnectivity with the greater community
- A walkable environment

Traditional neighborhood community design is perfect for the huge percentage of homebuyers seeking a diverse community as opposed to a monolithic 55+ age-qualified development.

The essentials of a traditional neighborhood development design include

- *Compact living:* Or higher density living, to facilitate a more walkable environment with the town center, village green, and other amenities within walking distance of more homes.
- *Town center:* A town center with as much commercial space as the community can support: coffee shops, convenience stores, restaurants, and other gathering places.
- *Village green:* A village green or large centrally located park for large gatherings and events. Smaller neighborhood parks, individually designed, can serve as identifying features for those neighborhoods.
- *Community pool:* Sometimes located near the village green.
- *Walkable streets:* Residential streets are narrower than typical subdivision streets and have walks separated from the street by tree-lawns or landscaped areas. Houses and their porches are close to the sidewalks to encourage interaction with passing neighbors.
- *Homes with rear-loaded garages:* This is done to remove garage doors and a plethora of curb cuts that interrupt walks.

Let's look at each element individually.

Compact Living

The homes in TNDs are a mix of two- or three-story single-family homes, townhomes, and condominiums. The multilevel buildings and homes use less land than the one-story homes preferred by the 55+ markets. The design concept of a two-story home with elevator for the active adult is one way to address this challenge. See Chap. 6 for details on these houses.

Another way is to increase the density of multifamily portions of the development and balance that with some single-story homes for the 55+ folks, but that brings up a marketing question. Is there enough demand for additional attached homes or multifamily homes? If not, the price of the larger lots required for the single-story homes must be raised or the additional land cost spread out to all homes in the community. Either way, it will result in higher home prices and a competitive disadvantage. With the two-story active adult homes concept, there will be no loss of density and no increase in price since the elevator cost is paid for in the savings realized by building two stories versus one. This same idea can be applied to the townhouses for the younger buyers in the TND community. An elevator makes carrying things—like groceries, from the garage at ground level to the kitchen on the second floor or hauling luggage for a trip from the third floor bedrooms down to the garage and back up again upon return—much easier for people of all ages.

Enclaves of homes for 55+ active adults can be included in a TND community just as they are in master-planned communities. Since TNDs don't have entry gates or perimeter fences, the enclave may be a block or a group of blocks instead of a gated neighborhood.

Amenities attractive to boomers and seniors can be located within their neighborhood of blocks or as part of the center for the entire community.

Town Center

A town center is ideal for 55+ homebuyers. Having a place to gather or just meet casually is essential for a rich quality of life, and having it within an easy walking distance is even better. A coffee shop is a great place for people to meet and talk before getting on with their day, or in the middle of it. A fitness center shared by the entire development works well in a diverse community. The younger people can use the facility early in the day before work and in the late afternoon when they return home, and the boomers can use it during the day.

Typically the issue with town centers is how to generate enough business for the commercial entities to stay and thrive. To help with this challenge, town centers are often not in the center of the community but on the edge near a heavily traveled road.

Locating a generationally mixed community on two sides of a major road can accomplish the dual benefit of allowing the commercial area to be in the center of the community and giving retailers and restaurateurs the exposure necessary to generate adequate traffic. In this scenario, pedestrian crossings above, below, or at grade can connect the residential areas by walk and trail systems. These crossing must be easy and pleasant to use or they can become obstacles rather than connections.

Village Green

A village green or park that is connected to individual homes by walks and trails is an important amenity. Boomers love to walk for exercise and to socialize. A park can be part of the open space that they so highly value.

The village green as an idea implies more than open space. These are gathering places for the community and icons of identity for a neighborhood or community. Often features such as bell towers, clocks, bandstands, or simply one ancient oak tree located in the center characterize a village green. Since the village green and the commercial area are gathering areas and a walkable community is the goal, these are places to group the highest concentration of homes. That way the most homes possible are within the shortest distance of these attractions, a philosophy that guides city planning.

Community Pool

A pool offers the opportunity for exercise, relaxing, and socializing. Pool decks are the setting for conversation, reading, card games, and people-watching. That's a community pool, but the 55+ people who enjoy water aerobics or walking in the water for their exercise prefer all-year access to an inside pool with a constant depth of approximately 3½ feet or a gentle slope of 3½ to 4½ feet on one side in the center.

The said amenities would seriously influence the 55+ homebuyer to move to a new community. The top six amenities desired by members of that market, according to Margaret Wylde's survey as detailed in her book *Boomers on the Horizon* are walking and jogging trails (52 percent), outdoor spaces (51 percent), open spaces (46 percent), public transportation (46 percent), a lake (37 percent), and an outdoor pool (30 percent).

Walkable Streets

Walkable streets provide pleasant sidewalks uninterrupted by driveways to front-loaded garages. The walks are separated from the street curb by several feet of landscaping, creating a feeling of safety and security from passing cars. Trees in these "tree-lawns" not only shade the walks but provide places to stop and chat with neighbors. All of this reinforces the walkable character of the entire community.

Rear-Loaded Garages

We've seen nothing in the research done on boomers indicating any strong preference for the garage location to be in the rear or the front. Placing the garage in the rear does clean up the streetscape, making the area more pedestrian-friendly, and the walks are easier to navigate without driveway interruptions.

Author Kephart visited a community in California sometime in the late 1980s. The lots and houses were narrow and projecting two-car garages nearly as wide as the homes hid the front elevations of homes. The walks attached to vertical street curbs made for annoying and difficult walking conditions. The driveways, 6 inches lower than the walks, created two steps on the city sidewalk, several feet of walk and two more steps into and up from the next driveway. His old photos were lost but the experience never left his mind. Since walking is the preferred physical exercise of 79 percent of the 55+ demographic, per the 2005 Del Webb boomers study, it would be logical to conclude that locating garages in the rear would make sense for boomers.

So, all in all, TNDs are a good choice for 55+ boomers unless:

- They don't want to live so close to other people.
- They don't want to socialize and meet new friends.
- They must have a one-story home and a large private yard.
- They want to live in a community with gates.

The diversity—including age—in the population of a typical TND is what the many, many boomers say they want, and the self-contained nature provided by the mix of uses in a TND creates that small-town atmosphere so valued by boomers.

Transit-Oriented Design

It's called transit-oriented design, transit-oriented development, or transportation-oriented development. In each case, the acronym is TOD (pronounced T-O-D, not Todd).

TOD is a planning concept intended to reduce American's dependence on the automobile and the massive freeway systems that clog cities and fail to deliver the convenience they promised. TOD is characterized by a large number of multifamily homes located within a short walk of a light rail or train station. Small stores, coffee shops, and restaurants are drawn to these centers, creating a small downtown connected to the major urban center by a mass transit system. For residents, well-planned TODs can provide everything they need without having to use their cars. Yet for the occasions where they want to drive, their cars are mere moments away by elevator.

> The TOD idea is in its infancy. Denver may be unique in that all of the area governments worked together since the beginning on the huge multicity T-REX rail and highway expansion project, and all of the governments are making plans around their stops. But TODs set up a hierarchy that TOD designers haven't fully grasped. It has everything to do with what happens between stations. Along with serving the residents that live within a 10-minute walk to the station, you need other reasons for people to come to that center: restaurants, movie theaters, services. Entertainment is a key component. If the draw is not there, people are going to go there just to travel and the economic dreams won't be realized.

Planning a TOD varies mightily from one city to another and even from one transit stop to another. What's appropriate at one station may not work at the next one. Maybe some stops should just be stops. If a TOD is to be, it can't rely on rail transit alone; buses must also stop there to draw more people than just the surrounding residents and the quick-stop kiss-and-ride crowd. Local demographics, geography, proximity of other stores and restaurants, and other local factors need to come into play in the planning process. The more everyone with a stake in the station and surrounding environs works together, the better the whole TOD will work for everyone.

> Even though they tend to be reluctant to participate in the planning process for TODs, do everything you can to get the people who run the rail lines involved. Hopefully that would avoid the fate of the Elburn, Illinois, rail station. Elburn is about 50 miles west of the Loop in downtown Chicago, and it's the last western stop on the Metra rail line.

They had big plans for a TOD center; it would be the lifeblood of the town. But it isn't connected to Elburn's downtown, which is just four or five blocks away, and people can't easily walk to it from town. The Metra rail people made it into a repair and storage yard for their trains; so today it's a lonely parking lot in the middle of a vacant farm.

Components of Transit-Oriented Design

- Rail or light rail station as anchor
- Area serves the pedestrian—walkable or navigable for all ages and abilities
- High-density residential development within 10-minute walk to transit station
- Mixed-use center within 10-minute walk, containing offices, retail, and civic services
- Limited structured parking within 10-minute walk from mixed-use center/transit station
- Ideally, bus service to and from rail station to bring more people, but not more cars to the retail and commercial areas

---

Many cities around the world have very sophisticated mass transit rail systems. Paris, London, Prague, and Vienna have systems so user-friendly that a nonnative language speaker can go anywhere they choose with only minimal taxi use. It is nearly impossible to use a car in Hong Kong unless you are wealthy, but with rail and ferries, a stranger can travel comfortably with the daily commuters. Cars are heavily taxed on usage in Singapore. Weekend or vacation use permits are more affordable, but daily usage is so costly that few choose to drive to work and for shopping.

Economics, to some degree, influence rail or other mass transit usage. Companies in Shanghai, China, have recognized the toll that commuting takes on their workers and some require them to live close to their office or work site, sometimes providing housing for executives.

Some of our own major cities in the United States also have far-reaching transit systems or are creating or expanding rail service, including New York; Boston; the tri-state area of Washington, D.C.,Virginia/Maryland; Los Angeles, Baltimore; and Denver. American cities can learn from their foreign counterparts and many are, but this is not the issue for a builder until plans are actually made and opportunities arise to develop a part of a new TOD district.

Denver may serve as an example for other cities, due to a major political accomplishment a few years ago. Nearly all metro area cities and counties, in an unprecedented demonstration of inter-governmental agreement, approved a region-wide light rail system. The combination of various kinds of new light and heavy rail train lines connect downtown Denver to the extreme southeast corner of the metro area and to the extreme northwest, nearly to Boulder, 40 miles away. The lines run to the extreme east to Denver International Airport and far west to the foothills of the Rocky Mountains.

The north/south line has been running for several years. The southeast line that opened in late 2006 has sparked numerous developments around its stations, largely characterized by higher-density condominiums, podium buildings and mid-rises. A few high-density, attached single-family homes have been built close to some stations too.

The one type of home you will not be building in a typical TOD is a one-story, single-family detached home. TODs may include multilevel townhouses, but few in existing developments have elevators. That's got to change for the baby boomers, even in the unlikely event that stiff joints, arthritis, or other limiting conditions don't plague residents in that generation.

One new multigenerational TOD the authors toured had many three-level townhouses, none with an elevator or any space to install an elevator. In fact, the stairs up to the entrance and between floors were rather steep. The authors think the developer missed out on a great opportunity to attract boomers who just plain don't want to deal with stairs on a daily basis.

In the next few years, residential elevators are destined to become standard in multistory homes, partly because of the aging population, partly because elevators make carrying in groceries and hauling luggage between floors so much easier, and partly because younger homebuyers increasingly are seeing elevators as something they want but don't need—something to brag about to their friends to show they can afford it. It's like garage-door openers were for boomers when they were growing up. Few people had them when older boomers were young, but within just a few years nearly every house on the block had one, fulfilling a want but not a need. Elevators in multistory townhouses in TODs, whether in age-qualified developments or not, will attract a wider variety of residents than those without elevators.

Most multifamily homes in a TOD, whether condos or apartments, are single-level served by a bank of elevators. Inside, the living space is wide open and the floor plans are similar to single-family homes. A balcony or terrace provides a direct connection to the outdoors much like a patio outside a house.

There are opportunities for small- and medium-sized builders in TOD districts. In many ways, they are simply master plans that contain a mix of uses and housing types. A master developer may create the plan and build the commercial areas or some of the housing, but smaller portions of these high-density areas, adjacent to existing neighborhoods, are planned for less intensive housing types to serve as a transition to the neighborhoods.

UNIVERSITY-AFFILIATED RETIREMENT COMMUNITIES

Retirement communities that are affiliated in some way with nearby colleges and universities are ripe for development for boomers, who pride themselves on being lifelong learners, staying young, keeping active, and getting involved in athletic and cultural activities. What better place is there to mix all of these characteristics than in a university atmosphere? Among those who agree are the developers of retirement centers near Notre Dame University, the University of Florida, Cornell, Purdue University,

Table 3-1

University-related active adult communities benefit both parties

Good for 55+ Residents	Good for the University
Continuing education opportunities	Fees paid for continuing education classes
Available teaching or assistant positions	Source of talent and experience to serve as teachers/assistants
Volunteering opportunities	Free or low-cost assistance by residents
Cultural events (plays, musical performances, etc.)	Audience for special events, paid and unpaid
Available state-of-art fitness centers	Source of income for use of shared facilities
A walking environment on campus for exercise and enjoyment	Potential for benefit from gifts and inheritances
Library access available within walking distance	55+ residents can provide a stable population on campus in off-season
Sporting events	Volunteers 55+ can help
Diversity of people and ideas	The experience brought by the 55+

Penn State, Stanford University, the University of Arizona, Lasell College, and Nova Southeastern University.

The numbers of these developments—variously called university-affiliated retirement communities (UARCs), university-related retirement communities (URRCs), university-linked retirement communities (ULRCs), and university-based retirement communities (UBRCs)—will be increasing and also will be "repositioning" or morphing to meet the surge of learning-hungry boomers as they reach their sixties. Andrew Carle, assistant professor at George Mason University, Fairfax, Virginia, and Director of GMU's Assisted Living/Senior Housing Administration, predicts that the current existing three dozen or so such communities in the next 20 years will grow to upward of 400, or 10 percent of the 4,000 colleges and universities in the United States. A handful of these projects got delayed or cancelled during the real estate market slump in 2008 and 2009, but many more are moving forward. "There are probably at least 100 universities that are at least looking at this, if not developing one," Carle said. The main reason is that the arrangement benefits both the retirees and the universities (see Table 3-1).

Some university-affiliated retirement communities are actually sponsored by universities, such as Lasell College in Auburndale, Massachusetts, near Boston. Others are completely independent from the university they're near, such as Classic Residence by Hyatt in Palo Alto, California, near Stanford University, with which there is no formal agreement. But the huge majority of such developments have at least minimal if not substantial ties with their proximate universities, sometimes to the extent of the university initiating the project, such as Lasell. There are many roads up the mountain, if your goal is to incorporate lifelong learning and the inherent advantages of living in or near a university atmosphere.

On each project, the more the developer and the university collaborate in the planning process, the more successful it will be for both. Although the politics of partnering with a university may be daunting, once the approval process is complete, the market position for a university-affiliated community is free of competition, ensuring a fair chance of financial success.

Education as a part of retirement may not be for everyone, but it is important to a significant number of boomers, particularly those more highly educated. A 2006 National Association of Realtors survey of nearly 2,000 boomers found that 40 percent of boomers had a degree from a college or university. Of the remaining 60 percent, over 56 percent of them had graduated high school and 17 percent of those had some college experience. A university allowing continuing education was ranked number 9 on a list of 25 preferred community amenities by respondents in Del Webb's Boomer Survey of 2005. Forty-six percent of those likely to move chose education out of the list that included walking, maintenance, gated community, golf, cards, and age restriction.

Living close to a university provides bountiful opportunities for boomers. A majority of residents of university-affiliated retirement communities have some tie to the university they are near. Universities offer residents special involvement in classrooms through volunteering, teaching, or being part of special class projects. They often give reductions on tuition fees for residents wanting to take classes. The libraries, research facilities, and fitness centers are top notch and entertainment is available, from concerts, lectures, and plays to sporting events of all kinds. The big sports of football, basketball, and baseball are well known, but universities also support swimming, track and field events, hockey, soccer, rugby, and lacrosse. Residents often get preferred or guaranteed access to events that normally sell out quickly and possibly discounts on tickets too, depending on the arrangement between the retirement community and the university. The sweetheart deal between Penn State and The Village at Penn State provides residents with Penn State ID cards that give them access to the university's libraries and campus events, including hard-to-procure football and basketball tickets and reduced fees on tennis and golf. They may take classes at no cost if space is available. Student interns go to The Village to help run the healthcare facility and teach daily exercise classes.

Another benefit to residents is the proximity of coffee shops and restaurants, as well as all of the necessary services from dry cleaning to shoe repair, facilitating walking rather than driving to those places. Campus plans are uninterrupted by streets or roads, making walking convenient, safe, and enjoyable. The old trees offer shade and the large, open spaces are sunny. There are benches for sitting, fountains to enjoy, and people to talk with. It's an ideal walking environment for a community.

Gerard Badler, managing director of Campus Continuum LLC in Newton, Massachusetts, consultants, and developers of active adult communities on or near college campuses, calls the university lifestyle an "academic country club."

Living in an environment energized by the pursuit of knowledge, debate, and the youth of the student body is inspiring. Active adults on campus or nearby are naturally infected by the atmosphere and encouraged to remain mentally and physically active.

The demand for residences near universities for people 55+ is already evident. Campus Continuum conducted a survey of 233 people aged 55 to 75. Although a sample of 233 people is not statistically significant, it gives us an idea of where people are leaning. Fifty-eight percent of those surveyed had a master's degree or higher. Of that group, 58 said they would like to live in a small college town; 37 percent said they prefer the suburbs; and 28 percent said they prefer a city. Over one-third said they would be interested in retiring to a location near their alma mater. Sixty-four percent said they were interested in volunteering on campus.

Existing University-Based Communities

University-affiliated retirement communities have been around for upward of 20 years, including those near Harvard, Duke, and the University of Arizona. Most existing communities of this type are continuing care retirement communities (CCRCs), oriented toward residents age 62 and up and include the full spectrum of living options, from independent living to assisted living to skilled nursing and dementia care. That model is sure to change with the surge of baby boomers, who don't want to wait to "retire" to live where and how they want to live and who see themselves as perpetually active, but for the moment, that is the reality.

One example is Oak Hammock in Gainesville, Florida, which is affiliated with the University of Florida, though it is not on campus, as are some developments. The 400 Oak Hammock community members have campus privileges such as access to performing arts events, libraries, research facilities, museums, and classes. Oak Hammock shares a 22,000-square-foot fitness center with the university, and the Colleges of Dentistry and Veterinary Medicine offer their services to residents. The College of Arts provides performance venues and the College of Journalism and Communications helps support high-tech services to the residential community. In fact 18 of the university's colleges have affiliations with Oak Hammock, and the university sponsored its creation.

Most university-affiliated retirement centers are CCRCs. Oak Hammock is and so is Lasell Village, a 210-unit complex in Auburndale, Massachusetts, near Lasell College in Newton. Some of them are a bit long in the tooth and are renovating to keep up with seniors' changing needs and desires. Meadowood, which is affiliated with Indiana University, at age 29 is updating its assisted-living units from the outmoded single-room-with-a-bath design, adding more space and making the units more environmentally friendly to the point of going for a LEED silver designation. University Place in West Lafayette, Indiana, which is affiliated with Purdue University, is planning to double the size of its independent living apartments from 600 to 1,200 square feet. And, Holy Cross Village at the University of Notre Dame, which also shares a campus with Holy Cross College, has expanded some of its units from about 700 to 1,600 square feet.

The line between active adult communities and CCRCs is beginning to blur. The authors could find no university-related active adult communities. However, some university-related CCRCs are beginning to add more independent living units, and they already have the huge draw of the university setting, which will draw boomers, especially those

Figure 3-6
Eight cottages at
Balfour lie at the
foot of the Rocky
Mountains. (*Photo
courtesy of DTJ
Community
Planning.*)

without children or other family to rely on to take care of them when they get frail or sick and need care. And, some active adult communities are starting to forge partnerships with nearby healthcare facilities and even CCRCs to be able to offer their residents the complete package.

New Generation of Communities

Independent living communities on or near college campuses are sure to proliferate in the next decade as activity-oriented, learning-hungry boomers gravitate toward them. There are very few now.

Balfour

One independent living community is located near the foot of the Rocky Mountains in Boulder County, Colorado, close to the University of Colorado campus (see Fig. 3-6). Balfour is a 162,500-square-foot independent living community containing 103 units, including 95 rental apartments and eight cottages. The primary design objective of DTJ Community Architecture and Planning was to create a sense of community for the residents.

The building's architecture is based on the design of a Colorado mountain lodge, with exposed wood beams and stone exteriors. Particular attention was paid to building massing, providing the appropriate level of detail and scale for the residents. The living room is a two-story grand space with monumental fireplace and exposed wood trusses.

The facilities include a large formal dining room, private dining, informal "pub" dining, offices, lap pool, exercise room, beauty salon, parking garage, and resident storage.

The opportunity to capture multiple views helped in locating the two- and three-story wings of the building. The site's sloping topography contributed to locating the pool/exercise facilities and parking garage at the lower east elevation for best access. The clock tower and community room are located in such a way as to create a point of reference for the general public.

A courtyard was used as a focal point to organize the many amenities. It also serves as an outdoor extension of the public living room, library, and bar, and provides a secure shared gathering space.

University of Montana at Missoula—A Cautionary Tale

The University of Montana at Missoula had vacant land available for development and searched for a developer willing to work with them to create a residential community that would benefit the university and the community residents. Oak Leaf Homes of Cincinnati, in partnership with a private developer and a prominent, well-respected alumnus of the university, made a proposal for a new community for the 50+ markets at the invitation of the university.

Author Kephart planned and designed a phased residential development on the university's south campus, currently being used as a nine-hole golf course, plus some university student housing and administration. Among other plans were those to provide a trail head, with parking, for hikers climbing Mount Sentinel—the south campus sits at its base—and to preserve a north/south and east/west fairway landing space for hang gliders coming off the mountain. A significant classroom expansion was included in the south campus master plan at the university's request.

Preliminary agreements were forged between the university and the developer to work out a shared-use plan for the fitness center and other university amenities and services. A land-lease agreement was drafted to allow the university to retain ownership of the land for possible future university uses. The 50+ buyers would own their homes, which they would buy from the builder. Buyers would also make land-lease payments to the university. Additional considerations would be given to buyers electing to make a gift to the university later in life. All gifts would be tax-deductible. Since many alumni were expected to return to retire at their old university, it was reasonable to assume that a good number of them would include the university in their wills.

The project is currently on hold due to continued resident resistance, a potentially serious financial blow to the university, which was counting on the land bank they had at the south campus site for funds to expand programs and renovate or replace buildings that have outlived their usefulness.

The real lesson in this experience is to never underestimate the power of neighbors and concerned citizens. The university had the authority to impose its will on this situation, but officials opted to seek community approval, as a good neighbor would. The project owners went way too far in the planning of the project, only to realize the development was not welcome on that site.

Building a university-linked community is like building on an infill site in the middle of an established neighborhood. You must work closely with the community.

Guidelines

George Mason University's Assisted Living/Senior Housing Administration has developed a list of five criteria that Carle says should be in place to successfully develop what he terms university-based retirement communities. They include

- Proximity—preferably within a mile—to key facilities on campus, such as classrooms and major sports and entertainment venues
- A formalized, documented program of integration between residents and the university that covers classes, services, and venues the residents may take advantage of and under what conditions, as well as how the university will involve the residents in research, volunteer assistance, paid work, etc.
- Offering the full gamut of residential options: independent living, assisted living, skilled nursing, and memory impairment care
- Formalized documentation of the financial relationship between the UBRC and the university, regarding the land, services, and possible revenue-sharing
- At least 10 percent of residents having some connection to the university—alumni, former faculty/staff, or families of same

COHOUSING

A cohousing community is a cross between an enclave, a village, and a commune. Homes each have their own bedrooms, living rooms, kitchens, etc., but are clustered around common or shared facilities and amenities, often including a mega home of sorts that includes a super kitchen so residents may take turns cooking for all and a dedicated playground just for children of residents of the cohousing community, some of whom may share childcare responsibilities. Cohousing is a relatively new way to tend to the time-tried tradition of people striving to house themselves with independence, dignity, safety, mutual concern, and fun.

Some kind of common interest or characteristics draw residents together; they want to be very close to others of the same ilk. Some cohousing communities that exist today focus on spirituality, aging in community, LOHAS (lifestyles of health and sustainability), and sexual preference. Due to the small size of cohousing communities, this form of development is very well suited for custom building companies and other smaller-volume builders.

Words from the Father of Cohousing

Architects, spouses, and coauthors Kathryn McCamant and Charles Durrett, of McCamant & Durrett Architects of Nevada City, California, wrote the original book on cohousing, *Cohousing: A Contemporary Approach to Housing Ourselves.* They are commonly credited with importing this community model from Northern Europe and modifying it over the last 20 years and sparking the cohousing movement in the United States. Durrett's latest book is *Senior Cohousing: A Community Approach to Independent Living.*

The authors believe that cohousing will notably grow in popularity in the near future. Charles Durrett and Marysia Miernowska have much to say about cohousing for seniors. The following sidebar is an abridged version of what they wrote for this book. See the App. A for the complete text.

Senior Cohousing: The Village Solution

by Charles Durrett and Marysia Miernowska

Cohousing is a grassroots movement that began in Europe in the early 1970s and grew directly out of people's dissatisfaction with existing housing choices. People were tired of the isolation and impracticality of conventional single-family houses and apartment living, and desired a more sociable home setting that had less of an impact on the environment. The first cohousing communities were formed to combine the anatomy of private dwellings with the advantages of community living. The developments varied in size, financing method, and ownership structure, but shared a consistent theme on how people could cooperate in a residential environment to create a stronger sense of community and to share common facilities.

Senior cohousing has become a socially and environmentally sustainable solution, allowing successful aging in place with a high quality of life.

Design Guidelines

Cohousing communities typically cluster private residences around common spaces and facilities, often preserving open space as well. The common house is the heart of the community, and it serves as a gathering place for group activities, meal preparation, common dining, and meetings. The size of the common house depends on the people in the community, the frequency of meals to be shared and the budget available. Common houses may include guest rooms, multipurpose/hobby space, a meditation or prayer room, a fitness center, etc. It's entirely up to the group to decide on the features to include.

Parking and roads are typically located on the perimeter of the property in order to leave the center of the community as a pedestrian environment. People strolling or rolling by can stop and chat with neighbors rather than enter their homes by car through the garage, as in typical neighborhoods. The common house often acts as a "gateway" from the parking area to the community, thus providing another opportunity to interact with neighbors getting mail or while seeing what is cooking for the evening's common dinner. Pedestrian streets or a series of courtyards provide the primary circulation to and from parking, the common house and the private homes.

Along the way, you will find "gathering nodes," which can be as simple as a picnic table surrounded by flowers where neighbors can meet to have breakfast on a

Saturday morning or share a beer in the afternoon. These gathering nodes are usually visible from about four or five surrounding homes.

Indeed, visibility is reflected in the site lines of the site plan, as well as in the layout of the private homes. Unlike typical suburban homes, the kitchen faces the community, thus providing a view from the kitchen sink. The living room, on the other hand, is located to the rear of the home, opening to a private backyard.

The front porch becomes as lively as in traditional neighborhoods; it's a great place for people-watching, thereby facilitating personal interaction with people outside the home.

Cohousing communities may include retail space, as at Frog Song in Cotati, California, and affordable homes may be mixed with market-rate homes, as at Silver Sage in Boulder, Colorado. The homes can be lofts in renovated industrial areas; they can be townhouses, condominiums, single-family detached; and they can include existing and new homes within an established neighborhood. Any building form will do as long as the community design incorporates the essential design elements: common areas both indoors and outdoors; a pedestrian environment leading to entries to the homes; socially and environmentally sustainable design; and a common house where meals and other events can be shared.

Indeed, the four main characteristics of cohousing are extensive common facilities, design that facilitates community, a participatory development process, and complete management by the residents. In a series of meetings, future residents work with the architect to design a neighborhood that reflects their values and goals.

Senior Cohousing

As we get older and our bodies and minds age, the activities we once took for granted aren't so easy anymore: the house becomes too big to maintain; a visit to the grocery store or doctor's office becomes a major expedition; and if we can't drive to a friend's house for a visit, then we stay at home alone all day. Many, if not most, seniors recognize the need to effectively take control of their own housing situation as they age. They dream of living in an affordable, safe, readily accessible neighborhood where people of all ages know and help each other. But how many of those kinds of housing choices actually exist?

Usually limited to those 55 years of age or over, senior cohousing takes the concepts of cohousing and modifies them according to the specific needs of seniors. The result is a cozy little village that invites involvement, cooperation, sharing, and friendship—a recreation of earlier times when community participation was viewed as an essential part of social, mental, and physical health.

In a nutshell, senior cohousing is similar to the multigenerational cohousing model, with the following modifications:

- Careful agreement among residents about co-care and its limits
- Design considerations appropriate for seniors
- Size limitations—a maximum of 30 living units, usually 15 to 25
- Senior-specific methods for creating the community

Co-care is a unique quality of senior cohousing. Co-care exists because of the relationships and the proximity that exist in cohousing neighborhoods. Basic co-care issues are resolved far before move-in, where future residents discuss issues of aging in place. Cohousing facilitates helping a neighbor where otherwise help must be hired. Often, residents will choose to build an affordable apartment above their common house for an onsite nurse.

Senior cohousing gives senior residents both the independence and the proximate community they need to age in place successfully.

Cohousing in Action

The developer that's created the most cohousing communities in the United States is Wonderland Hill Development Company, based in Boulder, Colorado. The firm has built 16 cohousing developments in Colorado (see Fig. 3-7), New Mexico, and California and has four more in various stages of development. Jim Leach founded both the development and homebuilding unit over 30 years ago, sold the homebuilding arm to concentrate on development, and, after finding that it was taking as much management time to work with the contractors as it would to do the construction in-house, is now back to acting as his own general contractor.

Wonderland Hill's Wild Sage community in Boulder is a family-oriented community encompassing all ages, including several families with children. Wild Sage has 34 homes on 16 acres on an infill site, a density of 21 homes per acre. Homes are grouped in two- and three-story townhomes. Nine permanently affordable homes and flats and four "Habitat for Humanity" homes are included in the mix. The complex was built with green, sustainable principles in mind. It has a shared radiant heating system and the roughed-in infrastructure for solar heating; it is near both open space and city conveniences; and it is close to public transportation. The community was completed in 2004.

Across the street from Wild Sage, Wonderland Hill recently built its first active adult community, Silver Sage Village, composed of 16 homes on one acre. Six permanently affordable homes are included. Silver Sage has a 5,000-square-foot, two-story clubhouse, or common house, with one elevator serving all second-floor homes via a second-level walkway on the sunny side of the central courtyard. Residents each own their home and are equal co-owners of the common house. The shared building includes a

Figure 3-7
Cohousing communities such as Highland Gardens in Denver, Colorado, promote interaction between people by their very design. (*Photo by Mike Kephart.*)

gourmet kitchen, a large indoor great room, a large outdoor deck facing the Rocky Mountains, a guest bedroom for visitors, a workshop, a garden shed, and bicycle storage. There is also a space that could be turned into a residence for a future on-site community nurse or caregiver. Charles Durrett is the architect; Bryan Bowen of Bryan Bowen Architects in Boulder is the project architect.

Establishing a cohousing community requires extensive input from the group of owner-buyers. They must invest in the community at the beginning and participate throughout the process. Then they must handle their own management after construction with a self-managed homeowner association. Generally there are no rules or restrictions for determining who will buy in a particular cohousing community. Potential buyers usually self-select their way into or out of a cohousing complex. Wonderland Development has a questionnaire for potential buyers that ask them to make their own decision about whether the cohousing lifestyle is for them. They answer only to themselves. The following list paraphrases some of the issues for people to ponder:

- I respect other spiritual paths and do not hold mine as the only one.
- I value a sense of community with others.
- I appreciate diversity in others and in the community.
- I am interested in trying new experiences.
- I want to continue learning.
- I am open to change in the community and in myself.
- I am willing to participate in community activities.

- I value the environment and take action to help.
- I want to help to create the community I will live in.
- I have or would like to have a regular spiritual practice.
- I would like to volunteer to help others.
- I try to be as physically active as possible.
- I am willing to explore the mysteries of aging and death with others.
- I am willing to develop my talents.

The way people answer these questions should inform them of whether or not this is the community for them.

To date, according to the Elder Cohousing Network, only four cohousing communities specifically built for people over 50 exist: Silver Sage in Boulder, Colorado, completed in 2008; ElderSpirit in Abingdon, Virginia, completed in 2006; Glacier Circle in Davis, California, completed in 2006; and Wolf Creek Lodge in Grass Valley, California, to be completed in 2009.

The first age-restricted cohousing project was Glacier Circle Senior Community in Davis, California. It consists of just eight homes and is part of a bigger neighborhood. Most had known each other for decades through their local Unitarian Universalist Church.

ElderSpirit in Abingdon, Virginia, was next. A nonprofit organization, the Federation of Communities in Service (FOCIS), formed a corporation for the project, Trailview Development Corporation, which built the 29-unit complex on 3.7 acres. Phase I was 25 homes—12 for rent and 13 for purchase. Phase II of the age-qualified community was the common house, which has four one-bedroom apartments for rent to low-income people. The architect was the Highlands Group, P.C., in Roanoke, Virginia. The construction is simple, economical, and traditional in style, as opposed to the contemporary styling of Silver Sage.

Dene Peterson, the founder, says the values upon which ElderSpirit was formed are spirituality, mutual support, simple lifestyle, and respect for the earth, arts, and recreation, healthcare during illness, and dying and mutual assistance.

Leach says he doesn't understand why other builders haven't embraced the cohousing market, unless it just looks too hard to learn for a relatively small-niche market. Leach and his company, which includes his daughter, Shari Leach, provide a full palette of services to cohousing groups. They conduct workshops to teach people how to form a group; facilitate group formation; provide project management, conflict resolution, and marketing; and offer legal and financial support to newly formed groups.

REFERENCES

1. "Baby Boomer Report, Annual Opinion Survey by Del Webb," Survey Results Summary, Pulte Homes Inc., 2003.
2. Cross, Tracy, Various data, Tracy Cross & Associates, Schaumberg, Ill., September 2009.

3. Frye, William H., *Mapping the Growth of Older America: Seniors and Boomers in the Early 21st Century*, The Brookings Institution, Washington, D.C., May 2007.

4. Foligno, D., "Field Guide to Marketing to the 50+ Population," National Association of Realtors Web site, www.realtor.org. Accessed July, 2009.

5. Schriener, Judy, "Retirement and Education Are Profitable Bedfellows," *Engineering News-Record*, July 2–9, 2007, pp. 18, 21, 22.

6. U.S. Census Bureau data from The 2009 Statistical Abstract, press releases on www.census.gov, and e-mails with public information office, 2009.

7. U.S. Department of Housing and Urban Development (HUD) data: The 1988 Amendments to the 1968 Fair Housing Act and the 1995 Housing for Older Persons Act.

8. Wylde, Margaret, *Boomers on the Horizon: Housing Preferences of the 55+ Market*, BuilderBooks (National Association of Home Builders), Washington, D.C., 2002.

9. Wylde, Margaret A., *Right House, Right Place, Right Time: Community and Lifestyle Preferences of the 45+ Housing Market*, BuilderBooks (National Association of Home Builders), Washington, D.C., 2008.

4 Aging in Place, Universal Design, Sustainability, and Building Green

"Design addresses itself to the need."

—Charles Eames

WHAT'S THE DIFFERENCE?

On the face of it, the title of this chapter is attempting to marry two ideas that share no common ground, but that could not be further from the truth. People living on their Social Security payments or on their pensions and retirement accounts are living on fixed incomes, and though their income never increases, their expenses, including their energy bills, go up every year.

Aging in place, as initially conceived, was a group of techniques used in home design, such as building wider doors, eliminating steps, and generally making a home more accessible and usable by people as they age. Universal design (UD) is a parallel idea that simply extends this concept to include people of all ages, health conditions, strengths, sizes, and mobility levels. Sustainability, also known as green building design or building green, focuses on energy efficiency, environmentally friendly materials, and making homes compatible with the environment around them.

The authors have combined all of these concepts into one chapter mainly because, although they may have different names and slightly different approaches, essentially they are synergistic and they have the same goal, namely to make a home more sustainable over a long period of time financially, environmentally, and socially. They are like a variety of vegetable soups. Some have carrots, celery, and green beans; others have carrots, potatoes, and corn; some have corn, leeks, green beans, carrots, and chicken. But they all are healthful, nutritious, life-sustaining—and hopefully delicious—vegetable soups.

Homes that fulfill the tenets of each of the concepts allow people of all ages and conditions to stay in their homes for as long as they want in a way that's comfortable, safe, and compatible with the environment, all of which add up to cost savings as a side benefit to the homeowner.

Although many builders see them all as separate and dispensable, they are quite compatible. "You can do universal design and green design too," commented Rebecca Stahr, founder of 50+ consultancy LifeSpring Environs Inc. in Atlanta, Georgia. "Homebuyers ask for them. They are not a passing fad."

GOING GREEN

Forget straw bale houses, old tires rammed full of earth, mud bricks you make onsite, or underground caverns. Building green is common sense and good design, coupled with exceptional execution of details during construction. Constant supervision and inspection by a skilled technician are called for to achieve the necessary level of quality in construction. This fact makes green building more difficult to accomplish for a large production company but some are already building 100 percent of their homes at some level of green now, including Engle Homes, a national builder, and more are sure to follow.

Green Goes Mainstream

Remember that baby boomers made up millions of the participants in the first Earth Day on April 22, 1970, and have been enlightening the world about the environment ever since. **Acceptance of the green movement is now considered not only mainstream but mandatory.** Societal pressures are intense against anyone who does anything that is perceived to harm our environment. That extends to energy, water, and waste within the home. So it should come as no surprise that boomers in study after study have indicated a willingness to pay for energy-efficient homes. Margaret Wylde in her book *Right House, Right Place, Right Time* said after surveying thousands of homebuyers and prospects age 45 and older: "...the most frequently preferred technology upgrade is an energy management system. Some 60 to 85 percent of survey participants have said they want such a system."

Wanting something and being willing to pay for it are not always the same. But, Wylde cited several unspecified studies indicating that homebuyers say they will pay $10 more per square foot for energy-saving features. She also reported that in one study 67 percent of prospective buyers said they would pay an additional $2,500 for a home with a heat recovery system, and 83 percent said they would pay an additional $12,000 for a home with a heat recovery system plus insulated exterior walls with R-50 insulation inside the home. In her early 2009 study of active adult homebuyers, however, Wylde noted some pushback. Factors related to conservation and green characteristics ranked lowest among the choices offered, all of which had to do with the attributes of the community, homes, and amenities, none of which were related to their individual home. That makes it more important than ever that designers incorporate green features into the homes as smoothly as room placement, lighting, kitchen décor, and other features so that, as with cars that come with energy-saving devices, it's all just part of the infrastructure of the house.

Concern for the environment continues to grow in the United States, although other priorities such as the conflicts in Iraq and Afghanistan, healthcare costs, and the recession that may dampen the momentum of the environmental wave. However, a short-term expenditure pays big dividends and many boomers will focus on that.

For example, if buyers are currently paying an average of $210 per month in energy costs for a total of $2,520 for a full year, the simplest of methods can achieve $1,000 in savings. Proper solar orientation, window placement, window shading, and the use of high-performance, low-E (for emissivity) glass—a hermetically sealed dual-pane system with argon or krypton gas between the panes—can save up to 40 percent alone, which is $1,000 with no extra cost except for window overhangs and higher quality glass. Add a few other good fundamental green procedures such as high-efficiency furnaces and air-conditioning equipment, thoroughly sealed ductwork, and a quality insulation and air barrier package, and the savings will far exceed $1,000 per year.

Green Pays Off

It is definitely in your financial favor to go green from the get-go. In a report published in October 2008, research conducted by McGraw-Hill Construction Analytics in conjunction with the National Association of Home Builders, 40 percent of homebuilders said that going green made it easier to market their homes; 16 percent said that going green made it much easier. Further, 60 percent of the homebuilders surveyed claimed that buyers would pay more for green homes, a 4 percent increase over responses in 2006. Possibly of greatest interest in the survey is that the previous number one reason that homebuyers gave in 2006 for going green, "doing the right thing," was surpassed in 2008 by "quality," indicating that buyers are buying green homes for performance and investment reasons as well as because they feel it is the right thing to do.

Let's look at each element to consider when building green.

Building Orientation

Orienting homes properly is a site planning issue that has emerged as the popularity for green design and energy efficiency has grown. The direction a home faces in relationship to the sun can save 20 to 40 percent of energy costs. "Orienting" includes

- Choosing which home elevation (long or short) to face to the south sun
- Deciding what percentage of glass areas to put on the north or south
- Placing overhangs or awnings above south-facing glass

It all depends on your location. For example, in Arizona and other Sunbelt locations that are "cooling dominant" in home energy use, limiting south-facing glass may be a good design choice. Locating outdoor rooms and patios on the north, more shaded north-facing areas makes them usable for most of the year. The opposite is true in Minneapolis and other northern states where the warmth of the sun is most welcome.

Solar orientation may come into conflict with the direction of available views. South-facing glass is great for passive heating but generally the quality of south views will be degraded when the bright sun is in your eyes. The photographers among you know this;

taking photos of people when you and the camera are facing the sun will leave their faces shaded beyond recognition.

If you are both developing and building, you can control your site plan to maximize the number of home sites that are passive solar-friendly. Developing sites that are 100 percent green-friendly is not the goal. Such a rigid design constraint can compromise other design considerations in community planning, such as variety in streetscapes, views that may face the wrong way, an efficient use of the buildable areas, and preservation of natural features. A good building orientation design not only considers both solar heat gain and views, but prevailing winds, slopes, soil conditions, drainage, and more. Landscaping can be used to block poor views or shade the home from the hot summer sun and the choices of plant materials can conserve water and save money. Drip irrigation and collection of rainwater for irrigation are two simple techniques to conserve water. More sophisticated systems are available but unnecessary at the beginning. Keep it simple! However, those of you in western states beware, as many states in that region will not allow the retention of rainwater. As water becomes more and more scarce, that restriction may spread to the eastern states as well.

Proper solar orientation varies from region to region and with local conditions. Generally in the north—heating-dominant locations where the energy use for heating is greater than the energy use for cooling—it is best to orient homes so a long, window-friendly wall (no garages or baths) faces south with as many windows as possible located in that south wall. Seventy-five percent of all window area on the south is a good goal, but there's a trade-off. With few windows on some elevations, the appearance of the home from the outside is greatly compromised, particularly with styles of the classic periods such as Georgian or colonial.

The positioning of a cul-de-sac can make a huge difference in how houses are oriented. For example, on a typical cul-de-sac, orienting it north to south or east to west will only allow two of eight homes to have the long dimension facing south (or north in the southern states) to get the benefit of energy conservation. But if the cul-de-sac were rotated 90 degrees, then six out of eight would work for energy-conserving design.

Window shading (overhangs) and low-E glass are used to prevent sun and heat entry into the home in summer. In southern climates, where the cooling cycle dominates, home design for energy conservation is about preventing heat gain from the sun. Large glass areas are ideally on the north and the managing of humidity grows in importance.

Energy-efficient green homes need not express green technology in their design but certain architectural styles are more accepting of the necessary roof overhangs. In order to work effectively, the overhangs above windows may need to extend 2 feet or more beyond the face of the wall. The nonsymmetrical placement of glass areas on the north or south can also be a problem with some architectural styles. Craftsman-styled homes, Victorian, contemporary, coastal, mountain, prairie school, and similar eclectic designs are easiest to work with. Formal styles such as Georgian/colonial, federal, or Tudor (no overhangs) are more difficult to incorporate passive solar design principles into without them looking odd or out of place.

A south-facing front elevation is one of the most difficult orientations to work with to accept the benefits of the sun without its negatives. Garage locations factor into that as well. A home with one window on the front elevation and two windows into the garage in front are not helpful for passive solar heat gain. You don't want the garage to block the sun from the house in the north, and that extra protection from heat gain in the south will save money on energy bills. Styles are expressed most strongly on front elevations and, for most lot configurations and home designs, the front is one of the narrowest elevations with a large percentage of the elevation consumed by a two- or three-car garage. South-facing front elevations on homes with rear-loaded garages are more adaptable to passive solar design. Garage locations require further consideration. You don't want the garage to block the sun from the house in the north, and that extra protection from heat gain in the south will save money on energy bills.

Floor plans can be designed to conserve more energy by locating the bedrooms, baths, closets, laundry, and other small/closed-in areas to allow the sun to penetrate the home through the open living areas or to block the sun's entry, depending on your location (see Fig. 4-1). There will be time to study these things later; the important thing is to get started and to start talking about green with your buyers.

Figure 4-1

Proper orientation regarding the sun can save as much as 40 percent of the energy lost. (*Drawing courtesy of KEPHART.*)

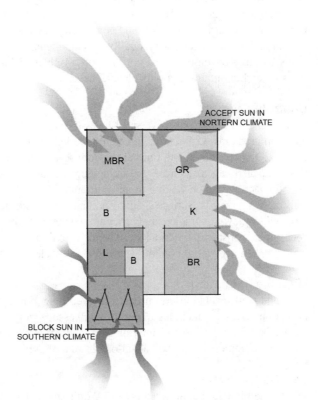

FLOOR PLAN ORIENTATION

Energy Conservation

Common sense guides 50+ buyers to understand the value of designing a home that uses less natural gas, oil, and electricity. In Margaret Wylde's study of housing preferences of age 55+ homebuyers or potential buyers, she found that they are willing to pay $6,380 on average up front when buying a home if it will save them $1,000 a year on heat and power.

"Saving $1,000 per year is easily achievable, without spending nearly as much as $6,380, and that's at today's energy rates," commented Paul Kriescher, principal of Lightly Treading Inc., Denver, consultants to builders and homeowners on green design and energy conservation. Costs for energy are rising as we approach the time that easily recovered gas, oil, and coal will be depleted. No informed 50+ homebuyer would expect current energy costs to go down or even stay the same.

In addition to good site orientation, saving energy can be accomplished with a few simple techniques, as outlined below.

Start with the Building Envelope

Consider life-cycle costs, not just construction costs, advises John Binder, project manager at KEPHART Community Planning Architecture, Denver. That includes evaluating innovative foundation systems, looking at wall systems other than traditional wood stud framing, evaluating HVAC systems by comparing initial expense with payback periods and system efficiencies, and considering advanced roof design, all for gaining maximum insulation, long-term quality, and maintenance reduction. "Evaluate the structure's building envelope based not only on a tight seal but also on how the structure breathes and maintains a comfort level," he said.

Control Leaks

Leaky windows, doors, and oversized holes through walls around pipes and wirings are common problems. Leaks through and around light fixtures, switches, and outlets are all culprits. Fireplace boxes and flues are seldom sealed with care and loose patches of insulation can also be leak sources. A simple blower door test (see Fig. 4-2) can locate leaks after the fact but the prevention of air infiltration is most effectively accomplished as the construction proceeds. Surface caulking around windows is never the fix for an improperly installed window. A complete air barrier around the roof, walls, and foundation is a simple concept, but constant inspections must be made to limit damage to this barrier during construction. You need a tight building envelope and sealed ducts as a base requirement.

Mind the Insulation

Roof and wall insulation are installed in every new home, but loose-fitting or sagging insulation leaves virtual holes where heat will escape from or enter the home. A sprayed-on insulation applied to the interior stud cavities after the electrical and mechanical trades are completed is a good all-in-one insulation, sealant, and air barrier.

Seal All Ductwork

What's the big deal about this anyway? Won't any leakage be into the home where we want the air to go? Yes, we want the warmed or cooled air to be delivered to all the rooms

Figure 4-2
A blower-door test can help to locate air leaks in the home. (*Photo by Mike Kephart.*)

in a home, but the problem with leaky ducts is that the air doesn't make it to the most remote rooms. It leaks into all the rooms along the way, leaving little at the end.

Also, ducts don't just leak into the places we want. Many ducts are located in attics, crawl spaces, or basements. When ducts leak into ventilated attics or crawl spaces, treated air and energy are lost to the atmosphere. Blown-in fiberglass insulation will not prevent this leakage even though it covers the ductwork. The home becomes depressurized inside when the return air volume pulled out of the home is greater than the volume of air that makes it to the spaces inside the home.

Leaky ducts can depressurize a home, causing outside air to be sucked in through holes and cracks in the building envelope and leaks around windows and doors. These leaks create cold spots, hot spots, and uncomfortable drafts in the home (see Fig. 4-3). Moisture carried in this air may also be retained inside walls, causing mold and rot.

Use Higher Efficiency Equipment
Ninety-two or 95 percent efficient furnaces, versus the old 60 percent models, not only use less energy to do the same job but they can be smaller in size and can be equipped with multiple-stage burners. This means less cycling on and off of noisy, high-volume fans. In addition, the lower-volume fans will run for longer periods of time, which creates

Figure 4-3

Leaky ducts lose heat to the outdoors while creating a low-pressure indoor environment that causes untreated outside air to be sucked in through small leaks everywhere. (*Drawing courtesy of KEPHART.*)

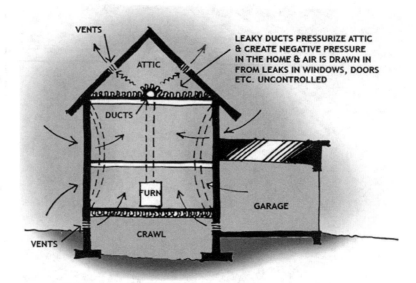

more even temperatures from room to room as well as less noise. Water heaters should be looked at the same way and on-demand heaters may make sense for your situation.

Temper Fresh Air

A tightly sealed home needs fresh air for the health and comfort of the occupants. Fresh air that is drawn into the return air stream can be filtered, heated, or cooled and the humidity can be regulated. One step better is a fresh air heat exchanger. This device heats incoming fresh air and exhausts cool, stale air. Minimal heat is lost in the process and the quantity and quality of fresh air are assured.

Reconsider Lighting

Compact fluorescent bulbs use a fraction of the energy of ordinary incandescent ones and they last several times longer. "Daylight" color 40-watt fluorescent tubes are great for indirect lighting, and they last for several years (see Fig. 4-4). Author Kephart estimated that the fluorescent lamps that have been in his home for 25 years have lasted for over 5 years between changes.

Dimmers turned low use less power and provide plenty of light for many activities like parties or quiet dinners. Well-placed reading lights or task lights at countertops, desks, and other places in the home can be put on separate switches to be able to turn them off when a lower light level is appropriate.

Improve Indoor Air Quality

Energy conservation by way of the ideas listed can not only save dollars, but will create a more comfortable indoor environment. Everyone, but especially 50+ homebuyers, can be very sensitive to temperature fluctuations, humidity, drafts, and overall comfort. Add problems with allergies to mold, pollen, or other substances and indoor air quality looms as an important consideration.

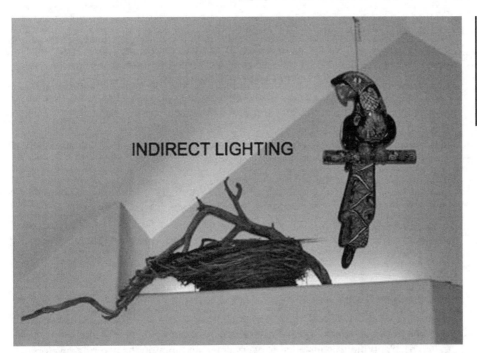

INDIRECT LIGHTING

Figure 4-4
Indirect fluorescent
lighting saves
energy and lamps
seldom need
changing. (*Photo
by Mike Kephart.*)

Further air treatment and filtering can help with molds and pollens, but unhealthy toxic vapors in a home cannot be easily filtered out. Eliminating their introduction into this closed environment is the best choice. Low-VOC (volatile organic compounds) paints, insulation, adhesives, and glues—without toxic chemicals such as formaldehyde or cyanide—and a safe substitute to conventional particle board will eliminate much of the indoor pollution. Carpets, another offender, can be aired out by the manufacturer to reduce off-gassing, or you can select nontoxic carpets. Even new furniture can be a source of indoor pollution.

Although a healthy indoor air quality does not create direct savings, it does help to build customer satisfaction.

Author Kephart described the process he and his wife went through to modernize and green up their home:

My wife and I saw the time coming for us to deal with our fixed income now that we are both retired. We elected to upgrade our home to reduce our energy consumption and our monthly energy bills. Lightly Treading Inc. analyzed our home, looking for places where we lost energy that could be cost-effectively improved. They conducted a blower-door test as the primary investigative tool to locate air leaks. In the test a canvas seal with a large fan in it was fitted to an open door of the house. All other doors and windows were closed as tightly as possible. The fan was turned on, sucking air out of the house, creating negative air pressure inside.

The technician then ignited smoke sticks and gave us each one. We followed him around the house, testing each light fixture, door, window, light switch, or other joints. We could actually feel big leaks as air rushed through from outside, and the smoke sticks detected smaller leaks and created some drama around the test. We could each verify each leak with our own stick and it wasn't long until we were satisfied that we had a very leaky house. We discussed possible solutions for each leak and computed the various costs, looking for cost-effective ways to reduce the leaks and conserve energy. The technician ran other tests to analyze the insulation properties of walls, windows, doors, and roofs.

We had two 30-year-old gas furnaces that were initially rated at 60 percent efficiency when they were new, and they had grown even less efficient as they aged. That was an easy call; we replaced both furnaces with new 95 percent efficient equipment, saving at least 35 percent of the gas we used to burn. Tankless water heaters were another easy call, though they didn't net as big a savings as the furnaces.

We insulated our old flat roof areas that were totally without insulation. Some 30 holes, each 3 inches in diameter, were drilled into our ceilings so the insulation could be blown in. A new layer of drywall on all of our ceilings was added to patch the holes. Caulking was installed at every crack and joint in the exterior envelope. We elected not to add exterior foam insulation and stucco over our old brick walls as recommended by Lightly Treading. That would have been very expensive and the return we could expect was marginal. We chose to leave our windows and doors alone, even though they also leaked air, as they were costly and we had replaced them a few years before.

Lastly, we added a solar photovoltaic array purported to replace 60 percent of our energy cost for electricity. The panels were placed on the north half of our flat roof to avoid the continual shade from our trees that help cool the house in summer. We had long ago remodeled the house using passive solar design principles by placing most of our windows on the south side and providing summer shade. Our house is a long, narrow shape with the long walls north and south. There are only two small windows on the north side and one on the west. We did place a bay window on the front, which faced east, even though that is not a recommended orientation for windows, as a concession to the conventional architecture in our neighborhood.

The results of our efforts are borne out in our reduced utility costs. Through the last two winters we have saved 25 to 30 percent of our gas consumption and well over 50 percent of electrical costs. Gas costs during summer have not changed more than 10 to 15 percent, which is nearly totally for water heating and a little cooking. We eat out most nights. In summer, the electrical savings is close to the as-advertised 60 percent. We are still tracking the charges. Other than the tankless water heaters, which are tricky to operate and, in our experience, give us only minimal savings, right now we are happy with the decisions we made.

Water Conservation

Drip irrigation and xeriscaping are two ways to conserve water outdoors. Minimizing grass areas and collecting and redistributing rainwater help as well, but sometimes a special permit to store rainwater, even for a short time, is required and may not be allowed by your city. Outdoor irrigation is the chief consumer of domestic water and once dealt with you can begin to look inside the home for ways to conserve there. Designers and builders are all familiar with low-flow shower heads and reduced-flush toilets since many codes require their use, but the water closet designs in Europe have seen past this. Toilets there have both a small flush or large flush option. The small flush uses less water than American toilets and the large flush does the job in one try. With our current reduced-flush toilets in the United States, we still must flush 2 or 3 times using even more water to handle the big jobs. American manufacturers are just now coming out with dual-flush toilets, although they are more common overseas.

Some people advocate the collection of water from shower and sink drains (grey water) to be used for irrigation. This requires a storage tank as well as a separate drainage pipe system. This is only for the more experienced among you. Bacterial growth inside the system can be a maintenance problem and the effectiveness of the system is highly dependent on the environmental commitment by the homeowners, over whom you have no control once they've moved in.

Maybe the best water conservation and energy-saving device is an owner's manual. This manual would not only include instructions on how to operate equipment, but could come with a training session on using the conservation features in the home. For instance, "Never let the shower run while shampooing or soaping up, just as you are instructed to do on boats or at some island resorts." However, some shower valves are difficult to use this way; you need a valve that turns the water on and off and a separate valve to adjust the water temperature. That way the flow can be turned on and off without affecting the temperature.

More Ways to Go Green

As you get more advanced in your practice of building green, you can incorporate additional systems and features, such as

- Solar water heating
- Photovoltaic electric generation
- Residential wind generators
- Geothermal heating and cooling (using the constant 55°F ± temperature of the earth beneath the home)

In addition, as a homebuilder, to be truly green there are several areas to work on under this general heading:

- Using recycled construction products or those with a significant percentage of recycled materials
- Reducing the amount of waste materials generated when building a home
- Recycling construction waste materials

Identifying products to use in construction that are made of recycled materials is a matter of continuous research. New products are introduced every day. Controlling waste during construction is a matter of design that allows the full use of 4 foot by 8 foot sheet materials without cutting; framing details that are more efficient, such as 24-inch stud spacing; lining up joists and studs to eliminate one top plate in walls; and corner framing designs that use fewer studs.

In 2004, the Home Builders Association of Metro Denver launched its new "Built Green" program (now "Built Green Colorado"). Since its inception in 1995, more than 30,000 new homes have been certified under the program. The five elements of green design, as defined in the program, are

1. Sustainable site planning

2. Water efficiency

3. Energy efficiency and renewable sources

4. Conservation of resources/materials

5. Indoor air quality

NAHB's Green Home Guidelines were released in 2005. These guidelines are intended for the use of local associations in developing their own certification program, as already exists in Denver, Austin, Texas, and other cities and counties. The more voluntary programs that are developed, the less likely communities will see a need to legislate the design of green homes. This is "offensive design," meaning seeing an unfulfilled human need and setting out to fill that need because it's the right thing to do. "Defensive design" is the pattern of the past; the human need grows so strong that governments force the private sector (builders) to do something about it. However, the perceived need to react quickly to a problem we've been creating for decades has caused states and communities to adopt rating systems of their own faster than the authors of national systems can keep up. California, for instance, in 2009 implemented the most stringent energy conservation requirements in the country. Hopefully, in a few short years, consumers will come to expect every home to be green and we will no longer separate green design from good design, any more than we now separate other elements of design, such as style, color, or a home's relationship to the street.

AGING IN PLACE AND UNIVERSAL DESIGN

Universal design enables people to age in place. Let's define our terms. The National Association of Home Builders defines aging in place as follows: "In plain English, aging in place means remaining in one's home safely, independently, and comfortably, regardless of age, income, or ability level. It means the pleasure of living in a familiar environment throughout one's maturing years, and the ability to enjoy the familiar daily rituals and the special events that enrich all our lives. It means the reassurance of being able to call a house a 'home' for a lifetime."

The concept of aging in place includes many things outside the scope of this book, such as health services, in-home care (a major subject on its own), home maintenance, transportation, and socialization. Universal design is a broader term referring to new construction as well as the renovation of existing homes. One good description of universal design, from Wikipedia, is, "Universal design strives to be a broad-spectrum solution that produces buildings, products, and environments that are usable and effective for everyone, not just people with disabilities. Moreover, it recognizes the importance of how things look."

Universal design (UD) is a design concept that recognizes, respects, values, and attempts to accommodate the broadest possible spectrum of human ability. UD goes beyond the accessible, adaptable, and barrier-free concepts of the past. It helps eliminate the need for special features and spaces, which are often stigmatizing, embarrassing, different looking, and usually more expensive. The UD concept emphasizes user-friendly design, aiming to accommodate the needs of people of all ages, sizes, and abilities, including the changes people experience during their lifetime. These features should be seamlessly integrated into a home's overall design.

Since most of the ideas contained in these concepts regarding the construction of new homes or the remodeling of existing homes are concerning accessibility or the ease of use of a home, we will use universal design to refer to both of these related subjects from this point on.

Homeowners and homebuilders have taken a long time to evolve in the arena of universal design. The first serious effort started not in homebuilding but in commercial building. The Americans with Disabilities Act of 1990 forced municipalities and private commercial building owners and builders to address the issues. The Americans with Disabilities Act Amendments Act of 2008, which went into effect January 1, 2009, broadened the definition of "disability" even further, and mandated even more people-friendly accommodations. Once homeowners and builders embrace the idea that universal design is so much broader, subtler, and more beautiful than ramps and grab bars, the idea becomes palatable and practical for all. Homebuilders can take advantage of the work that legislators and commercial builders have already done and "borrow" the guidelines that apply to their own residential projects.

Homebuyers are somewhat in the dark concerning universal design. So are some homebuilders! One homebuilder, Christine Fortenberry of Fortenberry Homes LLC near Atlanta, Georgia, who has been building for the 50+ market for many years, reported being in a meeting with a homeowner and another homebuilder. During the conversation about renovating the kitchen, Fortenberry brought up universal design and when the other contractor looked puzzled, she asked him if he knew what universal design was. She recalled, "He said, 'Well, yeah, I've heard of it. It's when dishwashers are all the same size and all of the other appliance [types] are all the same size. That's what you're talking about, right?'" She commented, "We've got a long way to go."

Homebuyers don't necessarily have *misconceptions* about universal design and accessibility; they have few conceptions at all! Margaret Wylde in *Right House, Right Place, Right Time* said, "Most 45+ households are interested in stepless entries (see Fig. 4-5), low or no thresholds, wider doorways, single-story living or two-story homes that are designed to accommodate an elevator if needed in the future, and at least one walk-in shower.

Figure 4-5
Zero-step entries
and doorways
are easier to use
for everyone.
(*Photo courtesy
of KEPHART.*)

Buyers' knowledge of an accessible home typically ends there. As a result, many buyers then opt for 10-foot ceilings, kitchen cabinets that reach to the sky, and light fixtures that require more than a step stool to change the bulbs."

Many homebuyers think they already know what they will need in their homes to live there into their senior years. Most are proven wrong. Boomers typically are in denial that their bodies will ever be anything less than able to move, reach, bend, and be strong. Maybe they'll be lucky and remain as they are now, but it's unlikely that they'll be 100 percent healthy forever. Even healthy people as they age may not realize it but their eyes gradually absorb less light, and the same objects that look light and bright at age 20 look dark and dim by age 70. They don't notice it because it happens very gradually. That's why people need more light to read with, for example, as they grow older.

A survey of 104 adults between ages 44 and 62 conducted late in 2008 for AARP indicated that more than one-third of boomer respondents plan to move from their current home when they retire. In an earlier AARP study of 1,016 adults between ages 45 and 64 in September 2008, four out of five (79 percent) said they want to stay in their current home as they age. A large majority of boomers, 85 percent, indicated they think they will be able to. In both groups surveyed, most prefer a one-story home with more conveniences to make them comfortable.

Although older Americans progressively are enjoying better health, many become disabled or limited by conditions such as arthritis or heart disease. In the 2000 census, 14 million people age 65 or older reported some physical limitations. Even though the percentage of those reporting a disability fell from 26.2 percent in 1982 to 19.7 percent in 1999 and probably will keep falling as health-conscious boomers reach 65, even 15 percent of 75 million is more than 11 million people over 65 with some kind of disability. To further fuel the need for universal design, the same report says that with divorce being on the rise and the succeeding generations behind boomers being smaller, there will be fewer children to take in and help disabled parents or grandparents. That would mean that more numbers of older people will want or need to stay in their homes for the rest of their lives or at least as long as possible. Universal design would help them do that.

Boomers are old enough to remember early homes made accessible for the "handicapped." They were ugly and expensive. Homes outfitted with obvious, intrusive grab bars that don't match the décor, outdoor ramps that are clearly add-ons, and accessories that look like an afterthought just scream "handicapped." Universal design doesn't do that; it just makes life easier for everyone: middle-age people as they grow older, younger people if they break an arm or leg, small children, and visitors who can't get around as easily as they once could.

One of the authors of this book had a total hip replacement shortly before turning in the manuscript and it wasn't the older of the two authors! She had an appreciation of the need for many of the elements described here that she didn't have the year before, and even though she hopes not to need them again, she was happy for the ones she had (no steps, wide doorways, walk-in shower, grab bars in the shower and near the toilet) and was painfully aware of the few that were lacking (chair-height toilet, raised platform for front-loading dryer, higher-level electrical outlets).

Getting Started

If you're not sure what you're trying to accomplish beyond just a hazy general concept, reviewing some specifics can give you a good grounding. The Center for Universal Design at North Carolina State University issued a list of seven principles of universal design in 1997 that are still considered to be the standard. See the following sidebar.

The Principles of Universal Design

PRINCIPLE ONE: Equitable Use
The design is useful and marketable to people with diverse abilities.

Guidelines:
1a. Provide the same means of use for all users: identical whenever possible, equivalent or not.

1b. Avoid segregating or stigmatizing any users.

1c. Provisions for privacy, security, and safety should be equally available to all users.

PRINCIPLE TWO: Flexibility in Use

The design accommodates a wide range of individual preferences and abilities.

Guidelines:

2a. Provide choice in methods of use.

2b. Accommodate right- or left-handed access and use.

2c. Facilitate the user's accuracy and precision.

2d. Provide adaptability to the user's pace.

PRINCIPLE THREE: Simple and Intuitive Use

Use of the design is easy to understand, regardless of the user's experience, knowledge, language skills, or current concentration level.

Guidelines:

3a. Eliminate unnecessary complexity.

3b. Be consistent with user expectations and intuition.

3c. Accommodate a wide range of literacy and language skills.

3d. Arrange information consistent with its importance.

3e. Provide effective prompting and feedback during and after task completion.

PRINCIPLE FOUR: Perceptible Information

The design communicates necessary information effectively to the user, regardless of ambient conditions or the user's sensory abilities.

Guidelines:

4a. Use different modes (pictorial, verbal, tactile) for redundant presentations of essential information.

4b. Provide adequate contrast between essential information and its surroundings.

4c. Maximize "legibility" of essential information.

4d. Differentiate elements in ways that can be described (i.e., make it easy to give instructions or directions).

4e. Provide compatibility with a variety of techniques or devices used by people with sensory limitations.

PRINCIPLE FIVE: Tolerance for Error

The design minimizes hazards and the adverse consequences of accidental or unintended actions.

Guidelines:

5a. Arrange elements to minimize hazards and errors: most used elements, most accessible; hazardous elements eliminated, isolated, or shielded.

5b. Provide warnings of hazards and errors.

5c. Provide fail safe features.

5d. Discourage unconscious action in tasks that require vigilance.

PRINCIPLE SIX: Low Physical Effort

The design can be used efficiently and comfortably and with a minimum of fatigue.

Guidelines:

6a. Allow user to maintain a neutral body position.

6b. Use reasonable operating forces.

6c. Minimize repetitive actions.

6d. Minimize sustained physical effort.

PRINCIPLE SEVEN: Size and Space for Approach and Use

Appropriate size and space is provided for approach, reach, manipulation, and use regardless of user's body size, posture, or mobility.

Guidelines:

7a. Provide a clear line of sight to important elements for any seated or standing user.

7b. Make reach to all components comfortable for any seated or standing user.

7c. Accommodate variations in hand and grip size.

7d. Provide adequate space for the user of assistive devices or personal assistance.

Please note that the Principles of Universal Design address only universally usable design, while the practice of design involves more than consideration for usability. Designers must also incorporate other considerations such as economic, engineering, cultural, gender, and environmental concerns in their design processes. These Principles offer designers guidance to better integrate features that meet the needs of as many users as possible.

Authors: Bettye Rose Connell, Mike Jones, Ron Mace, Jim Mueller, Abir Mullick, Elaine Ostroff, Jon Sanford, Ed Steinfeld, Molly Story, and Gregg Vanderheiden.

Copyright © 1997 NC State University, The Center for Universal Design

Options exist for the level of design you want to build to, depending upon the needs of the occupants. The four basic tiers are

1. *Visitability:* Accommodates visitors in wheelchairs or with limited mobility at one or more entrances from a driveway or sidewalk, and all of main floor or living spaces and at least a half bath on main floor

2. *Adaptable design:* Features or infrastructure for features that are not needed now but may be in the future

3. *Accessible design:* Barrier-free design specifically for people with disabilities

4. *Universal design:* Accommodates people of all ages, sizes, and physical limitations

> Note: The Americans with Disabilities Act (ADA), which mandates that buildings and commercial facilities be built or retrofitted to accommodate persons with disabilities, does not apply to private residences or to residences in general.

It is a common misconception that ADA applies to residential buildings or houses. However, the 1991 amendments to the Fair Housing Act do regulate the accessibility of residences in buildings of four units or more. No federal law requires accessibility of any kind in single-family homes, duplexes, or three-unit buildings, but some states and cities may have such regulations. The above misconception has cost hundreds of builders and developers millions of dollars in lawsuits filed by the federal justice department, so it's possibly worth it to you to get this straight before starting construction on a residential development, particularly one with multifamily buildings.

Elements of Universal Design

You can do the minimum or you can go hog wild. The features of universal design are virtually limitless. The more of them you include in the home, the more inclusive the home becomes for all to be able to live in. The more you incorporate them into the design, the less obvious they become, the less they shout "for the aged or handicapped," and the more appealing the home will be to a wider variety of people. After all, eliminating steps and creating a gently sloped entry isn't just for wheelchairs, walkers, and crutches. It makes it easier to use a baby stroller, get a piece of furniture delivered, and maneuver the wheeled luggage we all have now.

Not all of the design elements below are feasible for all homes. Take into consideration who will live in the home and who is likely to visit, and also project ahead as if they will be staying in the home another decade or two, and go from there. Who would be likely to buy it on a resale might also be a consideration.

General Guidelines/Examples

- At least one no-step entry into the home, at minimum, accessible from the sidewalk or driveway; ideally, *all* doors at least 36 inches wide and *all* room-to-room thresholds are flat
- An open floor plan with flexible space that can be used for any of several purposes
- Hallways 36 inches wide or wider
- Turning space of at least 5-foot diameter in each room when possible
- Slip-resistant flooring
- Lever door handles on all doors, easy-grip fixtures on drawers and plumbing
- Light switches located lower than 48 inches from the floor
- Electrical outlets located higher than 18 inches from the floor
- Electrical panel mounted 18 to 42 inches above ground in location that's accessible to all

- Grab bars in bathroom, or bathroom walls reinforced for future grab bars that will support at least a 300-pound load
- Spacious roll-in shower with a generous, sturdy, chair-height seat
- Chair-height toilets
- Generous lighting throughout the home—natural light, task lighting, and ambient lighting; also install dimmers where feasible for energy conservation and mood lighting
- Multiple heights of cabinets and counters for easy accessibility for a variety of heights and for those when sitting and standing
- Appliances that can be reached and operated from a standing or sitting position
- Windows that are easy to open
- One-story living space, master bedroom suite, and guest bedroom and bath
- In multistory homes, an elevator or infrastructure for one to go in later
- If there are stairs, they have a deep tread (where the foot goes) and a low riser (the depth between steps) and handrails on both sides

There are things you can do in each individual space or room. Some of them are detailed below. Also see Chap. 6 and 9 for more room-by-room suggestions for homes and townhomes.

Entries, Doors, and Hallways
The goals are to make the home accessible to all people as they navigate into it and through it. That starts with the entry path to the front door. No-step entries are ideal. Getting from the sidewalk or driveway on an upslope can be challenging, but going back downhill on a too-steep slope can be downright scary, advised independent living strategist Louis Tenenbaum, so he recommended a maximum slope ratio of 1:12 or 1 unit (inch, centimeter, foot, etc.) of rise for every 12 units of run. Handrails are only necessary if the slope rises more than 30 inches off the ground.

The entryway at the top of the slope into the front door (or side door if that is where the no-step entry is) should be a minimum of 5 feet by 5 feet, longer if a wheelchair-bound person must turn 180 degrees to get inside the door, says Tenenbaum. "The top area must be wide enough to handle a doorswing and the entrant. There should be a two-foot clearance on the latch (nonhinge) side of the door. Five feet by five feet might not be enough," he wrote in his blog, *Aging in Place Guide*. He further advised that if a no-step entry isn't feasible, provide handrails on both sides of the steps and possibly also from the stoop to the door, and, if possible, put in "walker steps" with 3- or 4-inch risers and 18- to 24-inch-deep treads.

Alternatives to an extra-wide front door might be double doors or the addition of a sidelight that is hinged instead of fixed and can be opened up when the home's residents need a wider space and locked into place when they don't. You may have to special-order it. The wider door area is handy not only for a wheelchair but also for unloading oversized items from the car or when having furniture or appliances delivered.

All doorways, if possible, should be at least 32 inches wide, 36 inches wide if the approach is at an angle. Hallways should be at least 42 inches wide, plus extra room adjacent to the side of the door handle for maneuvering room.

> Author Kephart and his wife once stayed at a convention hotel and were forced to occupy the "handicapped" room. They still use those terms in hotels. This experience provided him with some interesting snapshots that he still shares with friends and associates. The renovation of the room to make it accessible according to the regulations was laughable. Doors banged against each other in the hallway, the toilet compartment had been removed but what remained was industrial-looking at best, and the closet ended up only 4 feet high to make room for the television. The ironing board they still insisted on providing was on the floor on the far side of the bed, out of view upon entering, but in the way when getting into bed.

Windows

The most commonly used single-hung windows in homes present a design problem for those in wheelchairs or those without the strength necessary to raise this window the first inch or two to pull it out of its tight seal at the sill. This seal is designed in for weather tightness and to conserve energy. To get around this issue you might consider using at least two casement windows on opposite sides of the home. Casement windows fitted with large crank handles require less effort to open to allow the air to circulate through the home. Another method would be to install a couple of motor-operated awning windows instead. An added benefit of these is their added security. Casement windows, when open, are very easy to pull fully open from outside.

Kitchens

The variations available to make a kitchen accessible and workable for all are nearly infinite. While the ideal would be to know exactly who will be living there, that's not possible when building whole communities. When designing custom homes you can put countertops and cabinets at different heights for shorter homeowners than for taller ones, put in varying heights if someone was in a wheelchair, etc., but even knowing the general characteristics of those likely to buy will help. Baby boomers like to think they will always be completely healthy and able to reach, bend, stand, and turn levers and knobs forever, and if that is not the case, they certainly don't want to be reminded of that possibility. The design should incorporate universal design elements in a way that makes the kitchen space attractive and user-friendly, both, without sacrificing one for the other.

There may be times when elements that may be convenient when placed one way just won't be palatable to boomers. One brand new complex the authors visited was specifically for people age 62 and over, yet the microwave ovens were all mounted above the stove, making lifting heavy containers of food challenging and the microwaves themselves out of reach for those in wheelchairs. When we asked about that, our very knowledgeable tour guide told us that residents and prospects objected to a low-mounted microwave, seeing it as something for elderly or handicapped people, which they didn't consider themselves. They would rather live with that inconvenience than have something that constantly reminded them that they're "old." Boomers are in more denial than other

Figure 4-6
Kitchen cabinets that pull out easily are more convenient for all. (*Photo courtesy of Mary Jo Peterson, MJP Design.*)

generations, so it would be good to offer as many options as possible for placement of elements and heights of countertops, cabinets, and workspaces.

Just as with all aspects of building for boomers, versatility is the goal in kitchens. Kitchens are the gathering place for families and guests. It seems that no matter where you *plan* for guests to be, they end up in the kitchen. With so many ages, sizes, shapes, and capabilities of people using the same space, universal design provides the road map to make the kitchen convenient, comfortable, and safe for everyone.

Layout   Ideally, several people can be in the kitchen at once without interfering with each other's activities. Several activity centers—sink area, stove, refrigerator, food prep, eating, cleanup, and just hanging out—can be spread around the kitchen and not clumped into one or two major spaces. The open spaces between work areas should be at least 5 feet in diameter. Activity centers should be accessible to all, including people who are in wheelchairs or just seated. If someone wants to sit and "take a load off" while working in the kitchen, the space should be able to accommodate that without blocking the way for others. One simple way to provide the required flexibility is to plan for the removal of base cabinets at key points to provide knee space for those seated.

Stand-alone islands are a matter of preference. Some may have a food preparation area and a stovetop for cooking, whether the oven is below or mounted in a wall elsewhere. Some may have a sink or a double sink plus counter space. Some are eating spaces. And, some are undesignated for whatever the homeowners decide to use them for.

Some are all one level; some are staggered levels with a few inches' difference in height. Some have storage below, either in traditional cabinets, open shelves, or larger, unpartitioned spaces with doors. Homebuilders and residents are divided in their preferences, and regional customs may help you decide what approach to take here.

Just remember that kitchens should create an easy triangle for the most common tasks performed in the kitchen, with the three points being the refrigerator, cooktop, and sink.

> Author Kephart's opinion: The truth is that builders often second-guess their architects by hiring a kitchen designer. Sometimes this is for the better and sometimes other considerations outside the constraints of the kitchen may be compromised. The final responsibility falls to the builder, though, so it is wise to have a working knowledge of this subject.

Appliances   The types of appliances in a kitchen and their placement, if done right, can make the space very user-friendly for all. Manufacturers offer a staggeringly large array of kitchen appliances and increasingly are doing the work of figuring out the best all-around designs so you can do less of it. Think convenience, ease, and flexibility, and you'll do fine.

Refrigerators   Consider whether a refrigerator with a freezer on the bottom or side instead of the top might be more desirable. Side-by-side refrigerator-freezers provide equal access to both sides, they can often include running water and ice dispensers. The narrower doors take up less space when open, but be cognizant that those narrow doors must open completely in order to be able to pull out shelves and baskets as designed. Fridges with freezers on the bottom are especially user-friendly for people who are in and out of their refrigerator much more often than their freezer. Refrigerator door swing direction is important in single-door models but is usually reversible on the job. That is not so with microwaves. You must pick the door swing direction that makes the most sense and remember when you build the reverse of the same floor plan that the microwave door swing may need to change as well.

Ovens and Cooktops   A cook stove with knee space below will be more usable for more people but there again is the flag in the buyer's eyes when they think that a detail like this signals that old people live here. A cooktop with a flat surface makes it easier to move pots and skillets off and onto burners. Regardless of whether you make room for knee space, for maximum convenience separate the oven from the countertop-level cooking element. Burner controls should be mounted on the front rather than in the back. Burners that are staggered rather than two directly behind two are easier to use.

Pull-down oven doors are easier for everyone to use, especially when the oven is mounted high enough so that when the door is open flat, it doesn't hit the lap of a seated person. Side-opening oven doors may be easier to use, but they are not common and you need to make sure they open from the correct side in each application. In either case, if the middle rack of the oven is the same height as the stovetop and counter, lifting is minimized when placing items in the oven or taking them out. Of course, use countertop materials throughout the kitchen that will withstand high temperatures.

You might even rethink the whole idea of a traditional oven. If you know the homeowner, ask if they use a microwave oven or toaster oven more often than a conventional oven. If the oven won't be used, lose it or place it away from the most heavily trafficked part of the room so the space can be available for more frequent ongoing activities.

Dishwashers A raised dishwasher enables people to easily pack and unpack it, whether standing or sitting. It is best if the top rack is the same height as the countertop. Or, alternatively, if cost is less of a consideration, a drawer-style dishwasher saves space and is easily accessed by everyone, whether standing or seated. A common practice is to stack two drawer-style dishwashers so one can be used without the other to save water and energy. An alternative to stacking them would be to put one drawer-style dishwasher in one place, say next to the sink, and the other somewhere else, either on the other side of the sink or next to the sink on the island.

Sinks Choices abound for sinks and their placements. They can be single or double or in between with one deep sink and a smaller, shallower sink next to it. More bending and reaching are required for a deeper sink, less for a shallower sink, but for some people, there is no substitute for the capacity of a deep sink.

Sinks can be in an island or against the wall, often beneath a bright window. The double sinks or one-deeper, one-shallower can be together or in two different areas. Wherever they are, one faucet for hot water and another for cold is just not a good idea anymore. Single-level faucets are easier for anyone to use.

Cabinets and Workspaces No matter how big or small a kitchen is, homeowners love counter space and always want more. Don't skimp on counter space on both sides of sinks, stoves, the cooktop and the refrigerator, especially the side that opens if it is a conventional refrigerator. Counter edges should not be sharp or square; rounded edges are easier on the bodies that bump into or lean against them and the hands and arms that use them for leverage.

Think about creating knee space under the counters and sink. Again, that's not only for wheelchairs but for anyone who wants or needs to sit while engaged in tasks. One author has a perfectly healthy back but it is strained and aches when she stands to wash and dry dishes that can't go into the dishwasher. Being able to sit would eliminate even that minimal back strain.

Varying the heights of countertops provides flexibility so that virtually all homeowners will find space to do whatever they want to do. Lower counters with knee space can turn the zone into a food preparation area, an eating nook, a place to sit while talking on the phone and still watching the food cook, a desk or area for a desktop or laptop computer if it is against a wall with electrical outlets, or any other activities they want to do while seated. Higher counters are especially good for food preparation or other work done while standing.

If counter space is limited, pullout counters, shelves, and cutting boards add a lot more convenient workspace, or, if mounted beneath the microwave, a handy transition place for hot dishes. You may even consider adding a second and third in areas such as by the sink, by the cooktop, and by counter space that might be used as a desk. A rolling cart helps in moving dishes and pots around easily, so you may opt to leave more floor area, if possible to accommodate a cart.

A walk-in or reach-in pantry closet isn't accessible to all, so consider other storage options such as pullout pantries (see Fig. 4-6). Arthritis and other factors make it difficult to grasp knobs so it is important for the handles and knobs on drawers, shelves and cabinets to be C- or S-shaped or otherwise easy to grip.

Eating Spaces   The name of the game for boomers, as the authors keep saying, is flexibility, so it's good to offer homeowners options for their eating spaces. The days of just one kitchen table and just one dining room table are long behind us. Give boomer homeowners places to sit for a quick snack, places for the family to gather, places for two to sit and chat, informal places, and more formal places. In the kitchen, you can provide nearly all of those options, all except the formal place, which should be in one of the other flex-spaces around the kitchen or great/living room.

Let's assume you have put in varying-height countertops in the kitchen. To turn a higher counter space into an eating area, residents can add bar-height stools. If they prefer to eat at a lower counter space, they can add chairs or shorter stools. In addition, they may—or may not—want to put in a table and chairs of whatever size they prefer.

There are other ways to approach this as well. Wraparound counters create a feeling of flow in the kitchen, as well as a visual division of space, part of which can be used as an eating area. Or, a detached breakfast table, possibly with a fold-down section, can be kept snug against a cabinet, partially tucked under an open area beneath a countertop, or moved to different places in the kitchen, depending on the occasion and number of people sitting at it. It can also double as extra counter space when not being used as a dining table. Once again, flexibility rules the day.

Storage   Nothing is more frustrating than trying to put something into or get something out of a drawer that doesn't pull all the way out. Make sure that drawers and shelves extend fully. Adjustable shelves enable homeowners to tailor their shelf space to their own needs. One place author Schriener lived in had adjustable shelves but instead of the mounting pieces being just on the bottom so the shelves could be raised, angled, and slid out and back in at a different level, mounting pieces were also added to the tops of the shelves, clamping them down so they could only slide in and out. However, the front edges of the cabinets overlapped the shelves, so they could slide only about half an inch forward and backward. Therefore, the adjustable shelves weren't adjustable at all, and most of the highest shelves had nothing on them because they weren't accessible without a step stool. So to help your homeowners utilize all of their cabinet space the way they want to use it, make the shelves adjustable and make sure they extend fully.

One great example of how to maximize storage in a small space is a stateroom on a cruise ship. The tiny space that is home to passengers for several days holds an amazing amount of "stuff." Every square inch of space is either functional or made into storage: shelves, drawers, nooks, crannies, closets, and open spaces of all kinds, sizes, and shapes. Kitchens offer loads of opportunities to create storage space. Storage should be spread throughout the kitchen and not concentrated in one or two areas.

You can get quite creative when it comes to carving out spaces for storage. For example, consider putting a bookshelf or mini-bookcase at the end of an island or eating area. It could be handy for cookbooks, do-it-yourself books, manuals, knick-knacks, baskets, or files. And, if you put the oven somewhere besides right under the cooktop, you can put big drawers for pots and pans in that space.

The corners in cabinets are often difficult to work with. Reaching back all the way inside is hard for anyone, let alone people with any kind of limitation. Full-extension drawers, pull-out cabinets, and revolving shelves solve the corner dilemma.

Baths

The bathroom is possibly the room in the house that most needs to be a safe place for people of all ages, sizes, and abilities. We are vulnerable there. We are alone more often there than in any other room. We have less clothing on. We need our privacy. The needs and desires of the homebuyer are often hard to determine here. If two people are to occupy the house, they may differ as to what they consider important, where things should be, and how they will use them, yet they may be embarrassed to share those thoughts and opinions with each other, let alone with an outsider.

Every homebuyer sees something different when looking at the bath in a prospective home. For instance, consider the bathtub. She may see a tub that will give her wonderful, luxurious bubble baths or drain away the stresses and strains of the day. He may see a space-waster that he would prefer not even be there. So what they want in the design, features, and tub placement would be very different. Or, consider a toilet. He may prefer a private toilet compartment so he can take his time without tying up the whole bath area and catch up on the weekly magazine he only reads in the "library." She may feel claustrophobic in such a compartment or just prefer the openness of the toilet integrated into the entire bathroom space. A drawback of private toilet compartments is that they are not generally accessible to all. For example, one of the couple may be having twinges of pain in their knees and want the extra leverage of a cabinet next to the toilet to help them get down onto or raise up from the toilet, which would not be possible in a space that contains only the toilet.

Preferences aside, the goal is to make the bathroom the most easily accessible and person-friendly room in the house, for everyone. The great news is that manufacturers provide a magnificent array of beautifully designed fixtures and hardware for the bath, probably more than for any other room in the house.

General Considerations   The doorframe should be at least 32 inches wide for a wheelchair to fit through straight on. It should be 36 inches wide if a wheelchair would be entering at an angle.

A 5-foot-diameter space in the central area of the room needs to be left to accommodate wheelchairs, or extra long space to provide ample room for wheelchair users to enter and close the door behind them.

Flooring should be nonskid. Smaller tiles provide better traction than larger tiles. Visual separation of tile types or tile designs make it easier for people to see at night, under low-light conditions, or if their vision is diminished by age or any other factor.

While higher lavatory countertops are popular, they are not accessible to all. One solution is to choose countertops that vary a few inches in height.

Grab Bars   The presence of grab bars, particularly when institutional in design, is perhaps the primary signal to buyers about their potential frailty. For that reason the authors recommend that bathroom walls be reinforced for the later addition of grab bars if and when needed if they are not installed initially.

Grab bars should look like design elements that enhance the space. Industrial-looking bars scream "for the handicapped" and turn off potential buyers. Grab bars or reinforced spaces should be everywhere a person would be likely to need them: in the shower, around the tub, and next to the toilet, at the minimum.

If for whatever reason you choose not to install grab bars during construction, block out the space so bars can be easily added later that can support at least 300 pounds. Most books will tell you they should support 250 pounds, but with the population's obesity rate edging upward, why not add the additional 50-pound capacity? It will vastly increase the levels of safety and helpfulness of the bars. The last thing a 275-pound person wants to worry about is if their weight will dislodge a grab bar when they need it most, sending them crashing, injured, to the floor.

One idea to consider is to install decorative towel racks that meet the specifications of grab bars, thereby keeping the design in the forefront while giving the homeowners the option of how to use the bars. Even if the occupants of the house don't need grab bars, guests might, and this way the towel bar won't be in danger of getting loose or failing when someone puts weight on it.

Bathtubs and Showers   Two ideas are at play here. One is to give the homeowners safeguards regardless of their disabilities or infirmities—and also if they have none—without sacrificing beauty. The other is to provide flexibility for occupants for whatever they want to do.

Louis Tenenbaum, an independent living strategist in Potomac, Maryland, said in an article on universal design on the American Society of Interior Designers Web site:

"Take flexibility for example. Instead of defining each space for specific use, prepare the space for alternate use and multiple routines. Take something simple, like a built-in seat in the shower. What if one of the clients wants the controls to the left and the other to the right? What if they want to bathe the dog in there? Those alternatives call for a space that is less defined, into which one can stand, lean or put and remove a chair. A built-in shower seat defines how the space is used. An open shower allows the space to be used flexibly. A spacious, less-defined shower allows more maneuvering, alternative uses and varied routines. Two can use the space simultaneously, whether for a shared shower or by a client and caregiver. Flexibility is preparation for the unknown, good-guess prognosis."

The days of the small tub that also serves as a shower are long behind us. Separate tubs and showers are definitely the way to go. Some homeowners might prefer to completely dispense with the tub, but in the discussions the authors hear among Realtors and at homebuilding conferences, they recommend keeping it if there is room for it if for no other reason than it helps later when the homeowner eventually wants to sell the house.

In nearly every newly constructed house the authors have toured over the last few years, in their opinion, they haven't gotten the bathtub right. Among other problems, the tubs are difficult to get into and out of, and they don't accommodate anyone that can't get up or down easily. A walk-in bathtub is friendly to everyone. There are many available that have water-tight doors, high seats to eliminate the need to lift or lower body weight, and knobs that allow for easy water temperature adjustment. However, some see them as "for the handicapped."

An inexpensive fix if the tub is already there is a folding tub seat or transfer bench. This is a seat that folds down to sit at the level of the tub edge to allow the bather to swing their legs into the tub. It can also be used in the shower by mounting it to the shower wall.

A separate shower is really mandatory for accessibility, flexibility, and ease of use. Inside the shower, a deep chair-height bench provides comfort for everyone, whether they "need" it or not. A mistake the authors see in most showers is a too-low and too-narrow bench. If shower takers have to lower and raise their body weight when sitting, it defeats much of the purpose for having the seat there.

If the house is truly to be step-free, the last step in the house, literally the one at the door to the shower, needs to be eliminated. A roll-in shower with no door is at one end of the spectrum, but you can also fulfill the spirit of universal design with a variety of other designs. Another option is for a flat rollover curb, which allows the user with limited mobility to step into the shower or a wheelchair to be rolled in with minimal effort. Anyone who has broken a foot or ankle or injured a knee knows that even that one small step can be painful and difficult to navigate and it would make recovery so much easier if it were not there. Whichever way you choose to go, the drainage has to be right to work, whether that is accomplished with angled drainage or a water retainage system. Installing an automatic door-opener is one way to provide ease of entry without making it obvious.

Sinks, Vanities, Cabinets   For accessibility to the sink, some designers say to remove vanity cabinets and replace them with a wall-mounted lavatory, since pedestal-style sinks are usually too tall. The nonarchitect author disagrees. She hates pedestal-style sinks and appreciates vanity cabinets immensely. Where else do you put hairdryer, curling iron, hand mirror, gels, creams, and other detritus needed to get ready for the day? And that goes double if the area serves two people. Regardless, universal design asks that you make space in the sink/vanity area for legs in a wheelchair or on a vanity seat. Remember, it doesn't have to span the entire length of the vanity cabinet. Insulate hot water pipes under the sink to prevent burning.

As to sink and cabinet height, homebuilders disagree on the ideal. Some say it should be low (30 inches) to make it wheelchair-friendly. Others say it should be higher (36 to 42 inches) so it's less strain on the homeowners' backs. The authors' tours of new homes ran the gamut on this. One challenge is that the average height of men is 5 feet 10.2 inches and the average height of women is 5 feet 4.6 inches so what would be more comfortable for one would be more uncomfortable for the other. Consider your own local market and culture, and ask what your prospects prefer!

Toilet Area   Newly built homes currently on the market are split on how they handle the toilet area. Some are integrated into the main area of the bathroom; others have a separate, closed compartment with a door. A private toilet compartment is not accessible or easy to use unless it is much larger than the typical ones offered on the market.

Whichever way you choose, this is one area where you need to provide a way to assist a disabled or obese person in getting up and down from the toilet. Two-thirds of adults in the United States are either overweight or obese, according to the U.S. Department of Health and Human Services. About half of those, or one-third of all adults, are considered obese, and the percentages for both categories are rising each year, regardless of age, race, gender, geographic location, or education. In public restrooms the people who will pass up an open regular stall to wait for the larger one with grab bars designated for the disabled include the overweight and obese, so clearly the need is there, even if they won't volunteer that information to you.

Lighting

Older adults need up to 6 times as much light during the day to see the same things they could see at age 20, and 16 times as much light in low-light or dark conditions. As the authors said before, the changes are gradual so people don't notice them as they happen. Also, the lenses of the eyes yellow as they age, and are more sensitive to glare. Further, aging eyes are slower to adapt from light to dark conditions. The University of Illinois Eye & Ear Infirmary's online *Eye Digest* describes it this way: "In comparison to younger people, it is as though older persons were wearing medium-density sunglasses in bright light and extremely dark glasses in dim light."

Boomers are closer to 80 than 20 and their eyes need increasingly more light. To help them see better, the goal isn't to install bright spotlights all over the house; the goal is to make up with a variety of lighting sources for what the eye has lost as well as make the home comfortable for residents and visitors of all ages. So include options for better and more light in more places.

Natural Lighting   The more natural light you can design into a home, the better, other than in Sunbelt areas where more sunlight equals excessive heat. Lots of bright natural light makes the space look larger and cheerier. Daylight can come pouring in through a wall of low-threshold sliding glass doors on the side that faces a patio or backyard. A skylight also can brighten up an interior space or a room lacking ample light from windows.

If possible, put in windows on more than one wall of each room. It evens out the light better than if they're all on one wall.

Frosted or textured glass or patterned glass blocks can bring light in from the outside and still maintain privacy, such as in bathrooms, including in showers and over bathtubs.

Ambient Lighting   Ambient light is the available light that covers the whole room. Natural light normally makes up some of it, and the rest comes from various types of artificial lighting. Universal design calls for multiple light sources to provide adequate light for younger and older eyes alike.

Extra light should be added to light the way on stairs and in hallways. Motion sensors in hallways and on the way from the bedroom to the bathroom can turn on low-level light fixtures at night to provide safe passage and save energy. Light switches with illuminated toggles in select areas also can help with nighttime navigation.

Universal design also calls for minimizing glare. That involves keeping the eyes from getting hit with the direct view of the light source. So naked light bulbs are not a good idea, nor are chandeliers, which are basically another form of bare light bulbs. If you mount recessed "cans" or downlights in the ceiling make sure the fixture is deeply recessed to minimize direct exposure of the eyes to the bulb. In bathrooms, position fixtures on either side of the mirror at about eye level—or fluorescent tubes over the mirror. Use opaque covers or shielding boards to reduce glare, soften the light, and direct light upward and outward. Avoid positioning light fixtures directly over where the occupants would be standing when using the mirror to minimize harsh shadows on their faces.

Fluorescent lighting minimizes glare when compared to incandescent lighting and also has the added benefits of energy savings and tubes that last 10 to 20 times longer than incandescent bulbs.

Fluorescent tubes aren't like the old flickering, buzzing tubes. Technology has come a long way. Just don't buy cheap fluorescent bulbs. Specify compact fluorescent bulbs or fixtures with T8 fluorescent tubes with electronic ballast. Bulbs or tubes should have a color rendering index (CRI) of at least 80 and correlated color temperature (CCT) of 2,700 to 3,500 Kelvin throughout the home, depending on where you need more light; 5,000 Kelvin in windowless closets; and variable levels of 2,700 to 5,000 Kelvin in bathrooms.

In kitchens, fluorescent tube fixtures mounted on top of cabinets provide nice, glare-free lighting for the entire room when light bounces off the ceiling. If there is not adequate space above the cabinets, you can mount a fluorescent fixture in the middle of the room. An incandescent light fixture shouldn't go in the center of the ceiling—it creates glare and shadows and doesn't light the room evenly.

Dimmers in the light switches enable homeowners to vary the levels of incandescent light they want at any particular time. Note that dimmers are available for some compact fluorescent bulbs and some fluorescent tubes.

Task Lighting   Light should be directed toward the task, not toward the eyes. In kitchens, supplement ceiling fixtures in the center of the room with light fixtures over sinks, stoves, and countertops, and under cabinets. Position them to shine down on work areas, a little in front and to the side of where the person would typically stand or sit to perform the tasks.

Fluorescent tubes to illuminate countertops may be mounted on the underside of cabinets. To hide them from view, they should be 2 feet long with prismatic acrylic lenses and electronic ballasts. Use T8 tubes with CCT of 3,000 Kelvin. Choose something that will shield glare and diffuse light, such as a fluorescent light with an opaque cover.

Summary of Universal Design

Ideally everyone's home should be a place where they can stay, fully function, and be comfortable regardless of whether they are in a wheelchair or have diminished sight, hearing, or balance; arthritis; diabetes; or memory loss. Even healthy people as they age may not realize it but their eyes gradually absorb less light, and the same objects that look light and bright at age 20 look dark and dim by age 60. That's why people need more light to read with, for example, as they grow older.

BRINGING IT ALL TOGETHER

Momentum for anything green and environmentally friendly has been building since before the first Earth Day. Some 20 million people participated, including students at thousands of schools and universities, that is, baby boomers. At first "green" was associated with radical environmentalists, derogatorily labeled "tree-huggers." But over the years the principles got more and more assimilated into our culture, laws, and construction methods. Homebuyers were finally willing to pay extra money out of their own pockets to proactively help save and protect the environment in their own homes—at least before the recession in 2008 and 2009. As the nation recovers from that, 50+ homebuyers again will be looking to benefit from green features in their homes, and that will increase as time goes on. Their children and grandchildren have been receiving the message that going green is no longer optional since they were youngsters, so even if the 50+ buyers are not personally committed to the concept, societal pressure from their peers and their kids and grandkids will push them toward it. If there's anyone who doesn't want to be perceived as politically incorrect in terms of being insensitive to their environment, it's a baby boomer.

AARP's December 2007 *Focalyst Insight Report* on "Green Boomers" claims that 70 percent of boomers surveyed say that "they feel a sense of responsibility to make the world a better place ... [and, for them,] social consciousness is a prevailing mentality." So-called "green boomers"—those who deliberately choose to spend their money on products that are environmentally safe—now comprise 40 million boomers or 54 percent of today's boomers, according to the report.

You may be missing out on opportunities if you assume a level of interest and willingness to pay for universal design and green design that's below what your buyers actually will be willing to pay for, and you may also run the risk of including features in homes that buyers aren't willing to pay for. That's where research comes in.

For example, green building shows the greatest growth in the Pacific region, followed by the South Atlantic and Mountain regions. The region with the least growth is the East South Central region of the United States, according to the McGraw-Hill Construction/NAHB report. Energy conservation is where buyers can see real economic paybacks for going green. You always want to add into the mix information about your own local conditions, demographics, psychographics, and competition.

Universal design, on the other hand, has no apparent economic benefit to homebuyers so if they don't need it now they may be far more hesitant to pay for a potential future need. You won't find real numbers about the popularity of UD. The changing demographics are clear, though. The population is getting older and we can project a far greater need in the near future. However, it's a real challenge to sell health and mobility aids to a group of healthy, active people, especially one that doesn't want to deal with those things yet. As builders and designers of communities for boomers, though, we know that UD has great value to those living in our community as they age and find that their homes continue to be fully usable. So when marketing UD to baby boomers, it may be helpful to use age-neutral words and phrases such as: "ergonomically designed;" "homes flooded with sunlight;" "easy and convenient to use;" "kitchens designed for everyone: short, tall, strong, or not;" etc. Esprit Homes coined "First Floor Living" for their homes at One Cherry Lane, a phrase that says it all without mentioning age, steps, or other obvious red flags.

Clearly both universal design and sustainability should have always been at the foundation of home design. We just lost sight of these bigger issues by concentrating on the daily issues we are all faced with in the homebuilding business. For example, a detail that is in conflict with the universal accessibility into a home still lingers in the hearts of builders: the step up from the porch into the house. Many builders have experiences where water ran in under a door and caused damage to a home, damage they were obligated to fix at their own expense. Flooding is a more severe risk in some areas so houses there are raised even more, which means more barriers to accessibility without thought to simply building on higher ground, an age-old solution to this problem that we seem to have forgotten.

When speaking of historic methods of dealing with the heating and cooling of buildings, we also seem to have lost the concept of orienting homes to accept the warming sun. We lost this idea when the lot pattern of subdivisions controlled the shape of homes more than the environment did. Single-family lots are narrow and deep to limit the length of street per house, but then many of these lots have that street on the south side. Add a garage on the narrow front and there is no way to welcome the south sun, and west and east exposures are difficult to control due to the low angle of the sun from those directions.

Hopefully we are opening our eyes to the full range of design opportunities and constraints, maybe for the first time in human history. Indigenous people worked closely with the environment but didn't live long enough to have to deal with the issue of caring for their elders. Self-defense informed design above all else a few centuries ago, forcing people to gather behind walls or to hide underground, as in Cappadocia, which today is central Turkey. The sun never reached them in either of these cases and the frail were most vulnerable to the violence. For many centuries, hunger, disease, marauding bands, and war were far more important than the lives of our older adults or the smoky clouds hanging over our cities. In early America we thought of our wilderness as without limit and the throwaway lifestyle was born. That has been one of our defining characteristics until just a short decade or two ago. Now we have an opportunity to begin dealing with our planet and its occupants differently. It's not just an opportunity, it's a mandate. And it's the boomers who are demanding it.

REFERENCES

1. AARP Web site, www.aarp.org. Accessed March, July, August, September, 2009.
2. Americans with Disabilities Act of 1990, available on ADA Home Page, www.ada.gov. Accessed September, 2009.
3. Americans with Disabilities Act of 1990, The U.S. Equal Employment Opportunity Commission Web site, www.eeoc.gov/policy/ada.html. Accessed November, 2008.
4. ADA Amendments Act of 2008, The U.S. Equal Employment Opportunity Commission Web site, www.eeoc.gov/policy/adaaa.html. Accessed July, 2009.
5. "Fair Housing—It's Your Right," U.S. Department of Housing and Urban Development Web site, www.hud.gov/offices/fheo/FHLaws/yourrights.cfm. Accessed August, 2009.
6. Figueiro, Mariana Gross, "Lighting the Way: A Key to Independence," AARP Andrus Foundation/Lighting Research Center/Rensselaer Polytechnic Institute, 2001.
7. "Height Chart of Men and Women in Different Countries," *Disabled World*, October 13, 2008, available at www.disabled-world.com/artman/publish/height-chart.shtml. Accessed November 2008.
8. Koppen, Jean, *Effect of the Economy on Housing Choices*, AARP, Washington, D.C., February 2009.
9. "It's Good to Be Green: Socially Conscious Shopping Behaviors Among Boomers," "Green Boomers," *Focalyst Insight Report,* AARP Services Inc. and Millward Brown, Washington, D.C., December 2007.
10. NAHB Model Green Home Building Guidelines, sponsored by the National Association of Home Builders and The NAHB Research Center, Washington, D.C., 2005
11. Jordan, Wendy A., *Universal Design for the Home*, Quarry Books, Beverly, Mass., 2008.
12. "Statistics Related to Overweight and Obesity," U.S. Department of Health and Human Services, June 2007, available at http://win.niddk.nih.gov/publications/PDFs/stat904z.pdf. Accessed November, 2008.
13. Tenenbaum, Louis, "Getting Started," American Society of Interior Designers Web site, 2009, available at www.asid.org/designknowledge/aa/inplace/active/gettingstarted.htm. Accessed September 2009, plus concurrent phone interview with Tenenbaum.
14. Tenenbaum, Louis, "Aging in Place Remodeling #3—Entries," *Aging in Place Guide,* blog, June 19, 2009. Accessed September, 2009.
15. *The Eye Digest*, University of Illinois Eye & Ear Infirmary, June 2007, available at http://www.agingeye.net/visionbasics/theagingeye.php. Accessed September, 2009.
16. "The Green Home Builder: Navigating for Success in a Down Economy," report, McGraw-Hill Construction, October 2008.
17. "Universal Design," Wikipedia, available at http://en.wikipedia.org/wiki/Universal_design. Accessed September 2009.
18. "What is aging-in-place?" National Association of Home Builders Web site: http://www.nahb.org/generic.aspx?sectionID=686&genericContentID=9334. Accessed November 2008.
19. Wylde, Margaret, *Boomers on the Horizon: Housing Preferences of the 55+ Market*, BuilderBooks (National Association of Home Builders), Washington, D.C., 2002.
20. Wylde, Margaret A., "Active Adult Housing Prospects," National Association of Home Builders Webinar, September 2009.
21. Wylde, Margaret A., *Right House, Right Place, Right Time: Community and Lifestyle Preferences of the 45+ Housing Market,* BuilderBooks (National Association of Home Builders), Washington, D.C., 2008.

5

Technology Turns the Tide

"We live in a society exquisitely dependent on science and technology, in which hardly anyone knows anything about science and technology."

—Scientist Carl Sagan

NOT-SO-SIMPLE TECHNOLOGY

Unlike any generation before them, baby boomers have been tech savvy for decades. They grew up with stereos in their homes and cars that would blast a sane person out of the room. They wired up their own component music systems with 4-ft-high speakers, tuners, receivers, tape decks [in the days before compact discs (CDs)], and turntables. They witnessed and used the first personal computers (PCs); some owned the earliest IBM PCs or Apple's Lisa, the predecessor to the Macintosh. They have grabbed on to the Internet and now freely embarrass their kids and grandkids by showing up on social networking sites. They are armed with laptops, smart phones, global positioning system (GPS) devices, and electronic toys and games.

Boomers like luxury. They like convenience. They like sound and fury when they want it and peace and quiet when they're done. Boomers are picky and demanding about their technology. They want it to fit into and enhance their lifestyle, not drive their lifestyle. They like technology to be simple (see Fig. 5-1) but they're up for the challenge of figuring it out if it's not. They don't mind tackling a learning curve for new gadgets or software, but they want manufacturers to make their instructions easy to understand—for example, written by someone for whom English is the first language—and kept to one page. In an AARP study of "Boomers and Technology" released in October 2009, one participant complained, "My little digital camera, which fits in my pocket, came with an instruction manual that was bigger than the one that came with my Subaru."

Boomers may be a bit behind the younger generations in adopting technology, but once they do, they are eager to share and they enthusiastically educate their fellow boomers, the AARP study found. According to the study, "Everyone wants technology that adapts to their needs, of course, but boomers see themselves as more assertive about demanding it than their children seem to be. Rather than early adopters, then, call boomers sensible adopters, who aren't about to change the way they live to fit technology. . . . Yet when presented with leading-edge technology that directly addresses their needs—even products not yet widely available—boomers are enthusiastic. In other words, when a technology makes sense to them, boomers may become early adopters and help lead the way."

Sounds like boomers are ideal candidates for the latest and greatest in technological advances in their homes, doesn't it? They probably are. However, the construction industry isn't as good at technology as boomers are. "A big mistake builders make is to latch on to whatever the latest trend is and 2 years later it's completely outdated," said Paul Doherty, managing director of The Digit Group in Shanghai, China, and former vice president at K. Hovnanian Homes. For example, if a homeowner would like to have a projector either for work or for personal use, building a space and wiring for a projector might seem like a good idea, but pico projectors—pocket-sized microprojectors that can magnify and display pictures or video on a wall—could eliminate the need for elaborate projector systems in homes. The technology is already available in Asia and may soon become a common feature in the United States. Technology is also guaranteed to be built in to appliances that up to now have been controlled by buttons and knobs. For example, voice recognition that's already common in cell phones ("Call Bob" dials a preprogrammed number) could end up in ovens. One participant in the AARP study said, "I can't understand the GE oven manual that tells how you program the stove to start your roast at a certain time. I'd like to just be able to say: 'Hey oven, turn on at a certain time, then, turn off.'" Technology is more fickle than a high school kid with a crush. The latest device roars into popularity and quickly collapses into obsolescence. Most boomers have owned vinyl records, eight-track tapes (what a good idea that was!), cassettes, Beta-max tapes, VHS tapes, component stereo systems, Sony Walkmans, CDs, DVDs, Blu-ray Discs, MP3 players, and all manner of computer equipment. Boomers know how transient the latest electronic device can be, so few will probably want built-in technology. However, they will want infrastructure to support whatever great gadget, gizmo, or system that comes out. Builders think they need to figure out whatever that great gizmo is going to be in order to be ready for it, but "future proofing" is designed to reduce the need to do that. Future proofing involves putting in infrastructure that's broad enough and flexible enough to work in a variety of scenarios, most of which you have no way of anticipating.

There are no standards of "this is the way to do it for everybody" like there are standards for everything else in construction. The lack of reliable guidance for designing and installing technology infrastructure in homes is not really the fault of those in construction, at least not totally. Technology isn't so simple that even a caveman can do it, to borrow from a widely popular TV ad campaign. There isn't really even a tried-and-true process for it yet, like there is for every other activity in construction. Doors are standard sizes and shapes, as are windows, and so forth. There are no such standard sizes or procedures for technology infrastructure.

Figure 5-1

Monks in Tibet brew tea with the sun. Will technology evolve into simple solutions such as this? (*Photo by Mike Kephart.*)

In fact, there isn't even agreement as to who is responsible for putting that process together and which actual trade will design and/or install technology systems. Should it be in the lap of the electrical contractor? The cable company? The telephone company? Who will take responsibility for the process and create one that works for the builder and the homebuyer? This is an ongoing debate and it doesn't look like it will get figured out any time soon. That only adds to the confusion of an already confusing situation. Every home is a custom job. It's quite a challenge.

How does a contractor know how to and how much to future proof? Look to the tradespeople who do the hands-on work. Doherty explained it this way: "The trick is to understand where that fine line is between providing *enough* technology but not *too much*. That is learned slowly over time by the subcontractors, the people who do the work. It's like, how do you know you need three coats of paint? Why not one? Or ten? The painting subcontractor just knows; he's learned it over time."

Great Expectations

When a homebuyer moves into a house or a renter moves into an apartment, chances are that the water will run, the electricity will work, and the phone will easily connect. But that may be about all. Some newly constructed homes have the workings for cable TV or satellite hookup already in place but generally an installer has to come out and do that job. It is not a plug-and-play home. That's where some other countries are far more advanced than the United States, just as they have been with cell phones for at least the last decade. That includes China. Doherty moved into a 100-year-old house in

Shanghai and was using the wireless Internet immediately. There the Internet technology is just another utility that the builder or landlord is responsible for providing, just like water, power, and heat. "It's no longer a luxury," Doherty said. "It's expected. It's an 'of course,'" He believes that when homebuilders get it that including technology infrastructure in their homes is an "of course," it will change construction. Home shoppers will choose a place that's already wired over a place that's not. When that starts happening, which he thinks won't be for another couple of generations, it will change construction because real money will be involved.

If it will take a couple of generations for technology to become as mainstream as (other) utilities in American homes, how will impatient boomers cope? Probably they will do just as they always have, that is, do it themselves, with gadgets in inconvenient places or wires hanging out everywhere to make up for the lack of supportive infrastructure.

Opportunity Knocks

If you are a small- to medium-sized homebuilder, this presents a huge opportunity for you. The big production homebuilders are not usually willing to be pioneers in technology. They want to wait until someone else does it first, then when they see that it will pay off for them, they will begin to move. But by then you could be way ahead of them. The sooner you help out your homebuyers by including at least a minimal amount of technology support when compared to your competitors, the sooner it will begin paying off for you. Word will spread and you will start getting a reputation as being able to provide what your competitors do not. If you don't know where to start, educate yourself. Read the tech blogs; they are filled with information you will find helpful and insightful. The high-tech manufacturers such as Cisco, Intel, Hewlett-Packard (HP), and Oracle are all accessible and would love to talk to you to get you started. Software and hardware were all together originally but got separated in the 1980s and 1990s. They are starting to come back together now, as manufacturers are beginning to take a holistic approach to them again.

Keep in mind that there is a social aspect to technology that is actually more important than the computing aspect of it. For example, computers originally started out helping people "compute." Hence, the name. But now most people spend more time communicating on their computers than they do computing. In addition, software on the desktop is getting more and more rare and less and less necessary as subscription services are becoming more sophisticated and popular so people can just log on to the Internet, get the software they need, go to any computer, log on, and keep working. That all makes it even more compelling for you to jump into the technology waters.

COMMAND AND CONTROL

As homes get "smarter," they will do more in the way of actually running the house, monitoring its functions, and reporting back to the owner. The authors won't make the mistake of trying to go into detail here about every new control, gizmo, gadget, system, and monitoring device you should be putting in. But they will try to steer you in a direction in which you can keep going.

Some of the functions you may want to give your boomer homebuyer the ability to use and have control over are

- *Communication:* Wireless connections linking many functions within the home
- *Entertainment:* Sound, video, etc.
- *Lighting:* Indoor lighting, outdoor lighting, task lighting, mood lighting, emergency lights
- *Environment:* Air quality, heating, ventilation, air-conditioning
- *Protection:* Alarm arming and disarming, ability to set zones and functions and change them at will
- *Healthcare:* Monitoring and communication for people with health issues

One area where boomers will probably spark the most innovation and growth is that of healthcare. Some things are being implanted, tested, or conceived that sound like the stuff of James Bond movies, such as a system that determines the shape of the person in the room to decipher whether she is standing, sitting, or has fallen down, said Louis Tenenbaum, independent living specialist in Potomac, Maryland. Voice communication can confirm the seriousness of the situation to a person who is monitoring the system in a remote location. The system depends on an electronic grid in the ceiling that keeps track of every space in the house, a kind of GPS for the home, said Tenenbaum. Sensors in the floor can also alert someone if there's any unusual variation of patterns that might indicate a problem. The extreme of technology's use in healthcare is the toilet that will check a person's blood glucose level in the morning.

Boomers who are concerned about aging parents not living in the boomers' homes, whether across the street or across the country, will see the benefit of the technological advances in healthcare monitoring as well as home system monitoring. The author of the AARP study elaborates and cites participants' reactions: "The concept is simple: equip the senior's home with sensors that can monitor activity, from what time they get out of bed to how much they move around to whether they've taken daily medication. Some systems even include one- or two-way video communication. For a number of participants with aging parents, this was attractive. 'I would feel more secure if my mother could be protected and I could get a daily report. Sometimes parents say "I'm fine" when they're not.' Said another: 'If there had been a monitor to tell me she hadn't turned off the faucet, that would have saved me about $100,000 in my mother's apartment.'"

With most boomers, however, Tenenbaum said you will want to avoid the subject of healthcare and point out the spa-like advantages of the technologies available to them. For example, wouldn't the new homeowners love to be able to turn the shower on and immediately have it at their preferred temperature, or set the device to fill up the bathtub with water at a certain temperature at a certain time? "They want the information but they want to pretend it's not intrusive," he said.

TECH STEP-BY-STEP

Most installations of infrastructure to make electrical or electronic things work are far simpler than the systems described above. Still, mistakes are plentiful. Ric Johnson, of Johnson Construction Services, Waynesfield, Ohio, consented to provide readers of this book a step-by-step guidance for what to do to get it right when tackling the *ideal* technological and systems infrastructure in a home. These are steps to creating a what is currently a "top end" installation; these things are not often found in new or existing homes now, but the authors think they will be in the not-too-distant future. The following narrative is from author Schriener's extensive interview with him at the National Association of Home Builders' International Builders' Show. They are his words, albeit sometimes paraphrased.

The biggest mistakes contractors make are

- They don't put enough lighting in
- They don't make the air quality good enough (not enough ventilation)
- They don't control harvesting the sun from light sources
- They don't provide many entertainment options

"Most people think they're doing great if they put a phone and television jack in every room," Johnson said.

When Johnson works with a builder and his client, he offers this as his process:

We start with the builder and his client. We have the basic electric covered and then we talk to the client about lifestyle. "Do you like to listen to music? Where? Do you feel like you're going to live in the kitchen more? Do you want items in the great room?" Most of the time they want a lot of flexibility in the kitchen or the great room. We pull up that drawing and talk about how they are going to lay out their furniture. Lifestyle drives the plan. Then we can proceed.

Great Room

In the great room you usually have a fireplace. Make that the center of focus. Put accent lighting there, a couple of cans to highlight a picture or the mantle. With perimeter lighting you can have separate controls for the main light, a ceiling fan, and accompanying light. If the client has a sliding glass door or a big picture window, put blinds or some kind of draperies on that control so the client doesn't have to get up to do that. As we're putting furniture around the room, we want to make sure we have enough room for table lamps, lights, etc.

Sit in the room. Listen to music. Put speakers in the ceiling. You want the homeowner to plug in an iPod without having to fumble around with that. Ask how they use the television. In the fireplace area, a recessed box is good; put it in the fireplace wall above the fireplace. Make sure it has double duplex electrical outlet (four plug-ins), a cable drop (can be for satellite or local cable), a data drop (CAT5 or CAT6 for the Internet), and a telephone jack. If the client says they like movies, put in a control drop for a DVD player and digital video recorder (DVR) so they can control them.

Put a five-pad control piece on the wall with a remote infrared (IR), controls for the television, drapes, and light. Hit "entertain" and it sets off a certain level of lights, and there should be one for normal activities, one for cleaning, and one for reading.

Kitchen

There should be three areas of lighting: ambience, task, and general. Put more lights in than normal because you need more light in the kitchen. For task lighting, use halogen puck light (disc lights). Little halogens sit in the upper cabinet. Light up the area so shadow doesn't come over their heads. Then in the perimeter put in recessed cans (lights cut into the ceiling) and the fixture up in the drywall. Depending on the kitchen island, use peninsula pendant lights (which hang down), add color into that, preferably color coordinated with the tile. Those lights can be light-emitting diode (LED), compact, or fluorescent.

Put a speaker in the ceiling. Have a control on the wall with an IR receivable. It should be the same thing with the lights on a keypad control and a timer if you know you're going to be in the area. It's like the sun's going down and the lights are coming up. Take it from 40 to 80 percent light, and the homeowner can change it to 100 percent if they'd like but 80 percent is usually enough and it's good for energy conservation. Mount the microwave and also recess the can opener above the cabinet or in a drawer that has a pop-up. All of the plugs are in there too. Put an under-cabinet coffeemaker in, prewired to the faucet and electricity, and preset it for coffee. It's all about convenience and lifestyle.

Also put under the cabinet a drop-down video screen with cable and aim it for a view of the front door. Push a button and talk to whoever is there. Or make it a nanny cam.

Put in an item that shows you if the garage door is up or down: a wall plate with two lights: green is up, yellow down. The reason we do this in our family is that one of us tends to leave the garage door up. Get a deadbolt keyed to your finger for the front door, one made by Schlage or the equivalent.

Put a video camera in the driveway and hook it up through the DVR. You can read the license plate, and you can access the DVR and see who's there. Like the authors said, all of this is the ideal.

Bed and Bath

Put more than one switch by the door and a switch by either side of the bed. Separate the ceiling fan and make it a remote. Put phone jacks on both sides of the bed. Tie a flat-panel screen to the video system. Put the infrastructure in—a backbox like behind the fireplace but with more wires, an interior cabinet for nanny cams, cable, most have their own CCTVs so a drop-in for that. For the music system in the bedroom, have two speakers also controllable with a wallbox and remote. Put in smoke detectors and carbon monoxide monitor where the fireplace and heater are. Use a different sound for carbon monoxide versus the fire alarm. Put a path of lights 100 percent down the hallway to the exit.

In the closet use a fluorescent light that has a daybright light so you can see better as compared to a standard white light bulb.

In the master bath, put the electrical outlets inside the vanity in a control plug strip. Make it heat sensitive so it kicks off the circuit breaker. Have a receptacle along the wall, put vanity lights on a dimmer, not with a system but where you can actually control it. Put the automatic fan on a timer or motion sensor, not on a switch. The problem with the switch is you turn the switch off and the fan goes off; with a timer it runs for another 15 to 20 minutes and keeps the house more static so mold doesn't have a chance to grow.

You can put a 12- to 15-inch flat panel TV screen in the bathroom. An option with Kohler is to put the TV in the shower. About 2 percent of clients do that. They are extremely driven, have to have information, and are wealthy enough to have it.

The Rest of the House

In the dining room, over the eating area, put a chandelier on a control, the perimeter lights on a control, usually with no sound or video because you can install that in the kitchen area.

On the patio, two speakers go outside; put in some kind of lighting, possibly lights in the deck. Try to have no steps. If you have a ramp, install step-lights in the bottom of the railing. It's controlled but you can override it.

In the garage, just do the lighting. They usually don't have technology drops, though some people also like a TV there.

On the front porch, you need three or four sections of light. The edges of the house need perimeter lights, a big light on the porch, and wall-mounted lights on the garage. For ambient light, a 24- or 40-watt lamp goes there. All four corners of the house get motion lights on a separate circuit. Set them so that no animals—cats or dogs—set them off, though a deer will. The whole house should be monitored by some kind of home automation system.

At Minimum . . .

If the above is out of your price range or you consider it overkill, do the following. Whatever lighting is normally there, increase it by 30 percent. Instead of a surface-mount two-bulb fixture, put either a three-lamp fixture in the ceiling and two sconces or the same fixture and four sconces. You don't want to have any place in the room that's dark or perceived to be dark.

You need at least one complete technology drop in each room—for telephone, TV, CAT5 networking drop. Baby boomers like dimmer switches or some kind of lighting control system. Most of us want some form of security in the home. The minimum would be a small security system that arms and disarms.

The "minimum" shouldn't add more than $3,000 over what the builder is already doing and you can sell it for a lot more.

REFERENCES

1. Doherty, Paul, The Digit Group, Shanghai, China.
2. Johnson, Ric, Johnson Construction Services, Waynesfield, Ohio.
3. Lohr, Steve, "What Do Baby Boomers Want from Technology?" *New York Times*, December 7, 2009.
4. Rogers, Michael, "Boomers and Technology: An Extended Conversation," Report, AARP, Washington, D.C., October 2009.
5. Tenenbaum, Louis, Independent Living Specialist, Potomac, MD.

6

Single-Family Homes and Townhouses

"Home interprets heaven. Home is heaven for beginners."

—Social Reformer Charles H. Parkhurst

If those who design and build homes don't give buyers what they want in a new home and community, they will stay in their old home until they find something that fits the dream for the next adventure of their life. However, the community where the house resides dictates many of its elements. The house shape and vertical configuration will be different for a retirement resort community than it will be for an urban infill location. The garage may be in front in one instance and in the rear in the other, and other rooms will be in different locations as well. Figure 6-1 shows a floor plan that completely surrounds a central courtyard, which is useful in conditions that are less attractive outside. This chapter will try to illustrate just how houses, duplexes, or townhouses are designed for each unique community concept.

The recession may have put a halt on the rising popularity of the McMansion, at least temporarily. The size of the average new house being constructed decreased by nearly 300 square feet between the beginning and middle of 2008, from 2,629 to 2,343 square feet, according to the U.S. Census Bureau. They had been shooting up prior to that, up from 2,227 square feet in 2005. As a comparison, median size homes in 1973 were 1,525 square feet. The National Association of Home Builders also reported that 80 percent of builders surveyed said they were placing an emphasis on smaller, lower-priced homes.

Boomers don't like to sacrifice anything they don't have to but they will go along with the spirit of a trend to conserve, and that includes space. Custom home designer Jim Phelps, founder of the Carolina Design Group in Cornelius, North Carolina, saw his market go from over 4,000 to between 2,500 and 3,500 square feet in just a year. He said, "Baby boomers are selling out of the bigger homes and moving into the smaller homes. They just don't want that huge house anymore. The baby boomer that's moving down [in size] may be driving a bigger Mercedes that has a plush feel. Their home has to have that same plush feel too. We are able to get high-end luxury items for a smaller square footage and they love it."

Figure 6-1

Floor plan with interior courtyard for locations without view opportunities. (*Drawing courtesy of KEPHART.*)

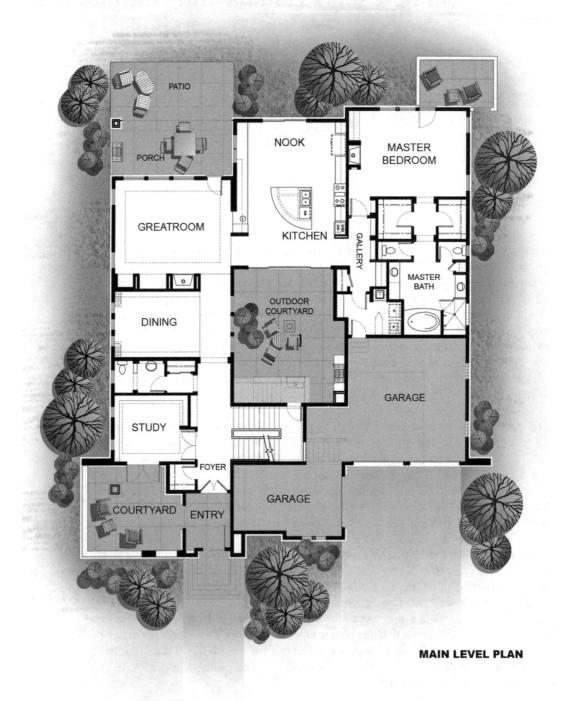

MAIN LEVEL PLAN

Designers and builders who focus on reducing the size of the homes they build without completely starting over on the floor plans are finding the boomer market receptive to homes that cut out fat rather than bone. Figures 6-2 and 6-3 show the same house in a before-and-after context. The "after" plan has fewer rooms and the remaining rooms do double duty rather than having a separate room for each function. Phelps said, "We just focus on the rooms they're actually going to use and get rid of the 'fluff' rooms. We take out the grandiose living room—that was all 'show.' We take out the colonnades and the hallways. We make the kitchen/dining/family room all one open space. We cut down on the square footage but you walk into a bigger space so the whole place feels bigger."

The key to making smaller homes palatable is keeping the quality high. "You use the luxury materials that these people have been used to," said Phelps. "If you miss that boat in this market, they're going to move on."

NEIGHBORHOOD-DRIVEN SIMILARITIES AND VARIATIONS

Context is probably the biggest factor when setting out to design single-family homes. Each type of neighborhood or community brings with it a set of expectations that are basic; the flexibility begins after those basics are fulfilled.

Old Is Now New

Most people in the United States don't want to be segregated by age, sexual orientation, or faith. This great American value is still held sacred by as many as 85 percent of people over 50 years old. This cutting edge idea, as it turns out, is also the oldest idea, harkening back to earlier times. However, there is a difference between the large communities of old and those of today. Today the homes in an age-mixed community are far more varied in design, with one-story homes for the 50+ crowd and compact two-story homes for the young families. Even major 50+ companies like Del Webb are building communities for families and active adults together. Del Webb, famous for their Sun City brand communities, now boasts their Anthem brand of mixed communities. Anthem Parkside & Country Club in Phoenix early on advertised with the slogan "Part Kid Kingdom and Part Adult Get-Away." All 10,000+ homes in this small city were open for sale to families and active adults but the single-story homes in the 3,000-home country club neighborhood were designed more for older adults than young families. The homes are all single-story/ranches, and they have their own amenity package within a defined enclave. There is no age or membership restriction on entry to this enclave and any resident in the community is free to use the facilities. The wave pool and water park in the other area are just more attractive to families with children. The residents select which community they feel more drawn to, and as a consequence one community tends to be composed of families, the other of older adults without children.

Figure 6-2

Before floor plans of home going from larger to smaller. (*Plans courtesy of Jim Phelps.*)

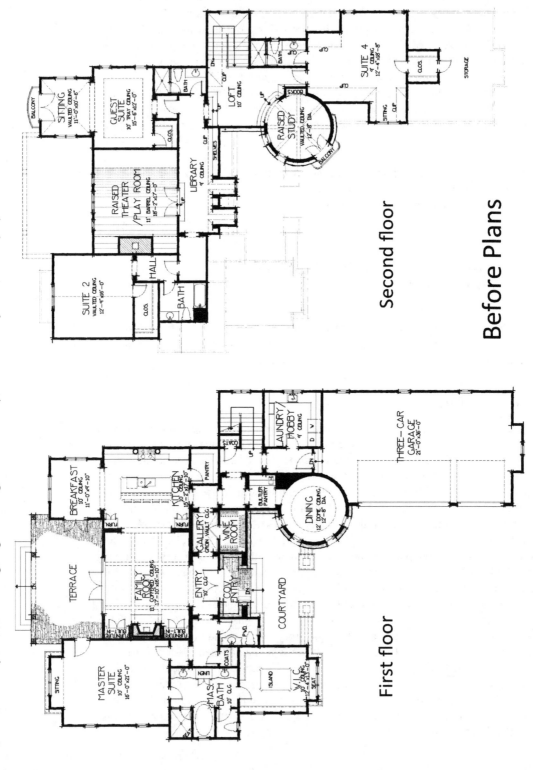

Second floor

First floor

Before Plans

| Figure 6-3

After floor plans of home going from larger to smaller. (*Plans courtesy of Jim Phelps.*)

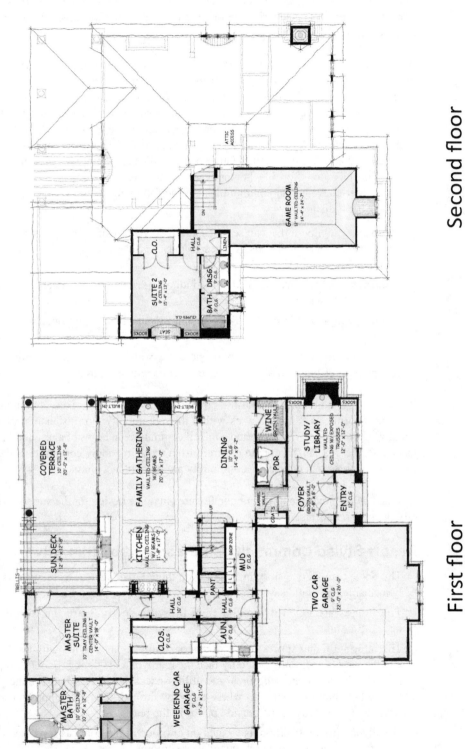

After Floor Plans

First floor

Second floor

In these types of communities, the special amenity package may be offered as a dues-required club but not necessarily age-qualified. The design of the amenities is sufficient to attract different types of residents.

If the community is large, a full complement of amenities will be provided for everyone, including clubhouses and pools. If the new community is smaller, you may need to rely on nearby city facilities, community centers, pools, and public golf courses. Be sure to check those places out when considering a small development. It's highly likely that any new community designed for all ages and close to home will be small. It's usually necessary to look beyond the city fringes to find land with room for a larger community that offers everything. Other communities are simply open to all, but with special considerations for 50+ homebuyers. These "new" communities may look the same as other conventional communities but the following design features and amenities make them distinctive:

- A separate clubhouse that can serve as a retreat when people need a little quiet time for conversation with friends, a place to play cards, relax, or exercise out of view from the 30-something tri-athletes.
- A separate pool with a constant depth—indoors, if possible—designed for water aerobics and lap swimming.
- Homes designed for the 50+ all on one level or with upper floors served by an elevator.
- No-step entries with connections to step-free walkways and other commonsense design features that will allow access to anyone with a disability.

Beyond the general concepts and the community amenities, a mixed community can take many forms, as can the homes within that community. For instance a traditional neighborhood design (TND) is often a mixed community in terms of ages, but the alley-loaded homes in a TND are quite different from those in a more conventional development, and homes in a cohousing community are particularly unique.

Outlined below are some of the unique features in home design for the various community planning concepts.

Resort-Styled Communities Personified by Sun City

- The homes are all one story in height even though the elevator technology is readily available and could save land and costs with two-story homes.
- The homes are generally wider than they are deep, as are the lots in these developments
- The informal living spaces (great rooms or family rooms), along with the kitchens and the master bedrooms generally face the rear while the dining room (used primarily after dark), second bedrooms, and utility spaces face forward next to the garage. Figure 6-4 shows a floor plan where the view through the home upon entering the front door is directly to the outside patios in the rear.
- Usually the master bedroom is located on one side of the home as separated as possible from the guest room and secondary bedrooms.

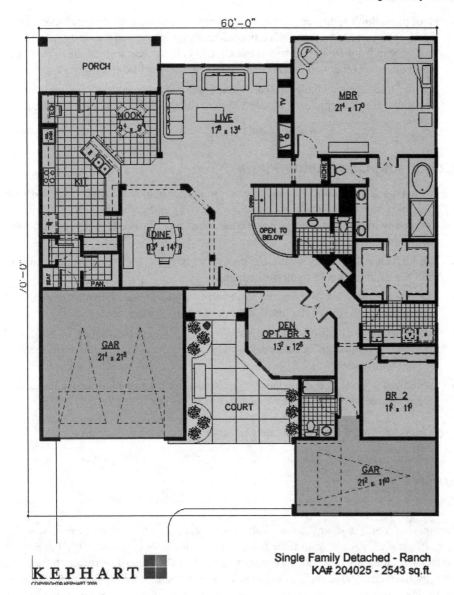

60'-0"

PORCH

NOOK
9⁴ x 9⁴

LIVE
17⁸ x 13⁴

MBR
21⁴ x 17⁰

KIT

DINE
13³ x 14²

OPEN TO
BELOW

NICHE

GAR
21⁴ x 21⁸

DEN
OPT. BR 3
13² x 12⁸

COURT

BR 2
14 x 11⁰

GAR
21² x 1¹⁰

KEPHART

Single Family Detached - Ranch
KA# 204025 - 2543 sq.ft.

Traditional Neighborhood Designs

- Homes in TNDs are primarily two stories tall or more, consistent with the TND concept of compact living.

- Steps up to front porches and inside remain a major contradiction but the creators of this concept have not yet been able to find ways to remove barriers in the form of stairs and resolve this dilemma.

- The detached garages, another traditional design element of a TND, are not always detached. This detail is particularly contested in northern climates where snow and ice linger in those spaces from garage to house during long winters.

- Floor plans don't follow as rigid a formula as found in the resort-style communities, but some things are common. One master bedroom is often on the first floor, which is usually in front facing the street; the kitchen is in the rear nearest the garage, and most bedrooms other than masters are on the second or third floors.

- In order to create a compact walkable community, houses are very narrow using as little street length as possible for each house. The multiple floors exist for the same reason.

Cohousing Community Plans

- In the typically very small cohousing communities, homes are often designed with two fronts, one facing the rear pedestrian walkway and gathering spaces where the community interaction takes place, and the other, more formal front facing out to the community at large.

- The kitchen is invariably facing the community activity on the common walkway from the common house to all of the homes.

- The living room is on the opposite front and is used for more formal occasions and to have some privacy for the individuals or the family when needed.

- The typically small homes may be in many forms: single-family homes, duplexes, townhouses, or stacked condominiums. The home forms are totally subservient to the community idea.

- Sometimes special accommodations are made for people who need an attached garage or access for a wheelchair, or other physical limitations.

See Chap. 3 for more information on cohousing.

WHAT PEOPLE WANT

Distinctive features in any home make it the place in which homeowners want to live. In general, people want their homes to provide a safe, comfortable haven and reflect who they are. The authors want to give you the benefit of research that has already been conducted, but, as always, keep in mind that this is *national* research on trends and preferences and local data is at least as important, if not more so, for you to make sure your ideas fit into the local and regional areas where you will be building homes.

The *Ideal* Home

Although markets vary, the following are the "top 10" attributes potential home-buyers have listed most frequently when given a blank sheet of paper and asked to describe their ideal home:

1. Large kitchen

2. Open floor plan

3. Attached garage with direct entry into the house

4. Low or no maintenance

5. Large closets

6. Leasing natural light

7. Gas (preferred) or wood-burning fireplace

8. Energy-efficient design, construction, appliances, systems, and features

9. Large bedroom, bathroom, and closet as part of an owner's suite

10. Outdoor living spaces, such as a porch, deck, balcony, or patio

Reprinted with permission from Right Home, Right Place, Right Time by Margaret A. Wylde

Margaret Wylde, a highly respected researcher of the mature market, is the founder and president of ProMatura Group LLC based in Oxford, Mississippi. Her extensive, detailed studies of the preferences of age 50+ homebuyers can provide invaluable insight into what home shoppers are thinking and wanting. She always emphasizes the necessity for local research because regional and local cultures and preferences can go 180 degrees away from what the national research indicates and you don't want to make the mistake of automatically taking national data as gospel in your own locale.

That said, ProMatura Group's 2009 study of active adult homebuyers revealed a lot of detailed information that can help you get grounded in the realm of building for the active adult and other 50+ home shoppers. The number one attribute of the home that influenced the buying decision in the study was the floor plan. Interestingly, the kitchen was at the bottom of that list (see Table 6-1).

PRECONCEPTIONS AND PREJUDICES

Architects and planners have their own preconceptions, which may or may not mesh with those of potential homebuyers. Homebuilders all spend a great deal of money and time on the research of markets, trying to get into the buyers' heads but almost no time looking at themselves and how their own experience and predispositions may alter the outcome as much as the research. In order to get a sense of the potential of this influence on design, author Kephart, the architect, interviewed five fellow architects and asked them a few simple questions. He wanted them to consider the fact that they may carry ideas in their heads, that they are looking for a place in which to try them, and that this preconception may play out to the improvement—or detriment—of the design of the community. He also asked them

Table 6-1

Things About a
House That
Influences You
to Buy

Floor plan, layout	21.7%
Size	15.4%
One level	11.6%
Location	10.7%
Price	6.8%
Open floor plan, great room	6.1%
Close to family	3.7%
Appearance, design, style of home	3.1%
Liked everything, had everything, fits needs	3.1%
Quality construction	2.6%
Lot, corner lot, yard	1.5%
View	1.3%
Upgrades, options, customization	1.3%
Convenience, less maintenance, no yard work	1.1%
Garage	0.9%
Basement	0.9%
Windows, bright	0.9%
Handicap accessible, wide doors	0.7%
Master bedroom on first floor	0.7%
High ceilings	0.4%
Landscaping	0.4%
Kitchen	0.4%
Available immediately	0.4%
Other	4.0%

to take him through their own personal process of design from what they see as step one through to the end. He tried to uncover how much these individuals are following the formal design process sanctioned by their profession, versus drifting off the set course for their own reasons. He promised them anonymity so they could speak freely.

Architect #1 felt that he always began the design of a community free of any preconceptions, but he revealed a clue when he admitted that certain parts of his work were "the fun parts." Would this lead him to spend less time on the parts that weren't fun? In follow-up questioning he indicated that it was a fixed "formula" for design that was held to by a client that made some things less fun. This was an early indication of the most important truth that came out in all five interviews. This professional felt that it was in the design of the amenities particularly where he was free to express the character of the community without restraint. The amenities in a development consist first and foremost of the clubhouse or community building, the functions in that building, and the walks and trails that connect the community. The other area that isn't constrained by the developer's "formula" is in the planning of the living units. It is there, as well as in the amenities, that he could fully influence the character of the community. He also spoke more about "invisible amenities," a phrase he brought up in his conversation. He characterized things such as the quality of materials and construction, energy efficiency, indoor air quality, and other features of green design, as "invisible amenities."

This list could include other benefits of good design that are also invisible:

- Eliminating unnecessary corners and height changes
- Designing dimensions in 2- or 4-feet modules to conserve materials
- Consolidating roof pitches for trusses to a few variations between models
- Organizing windows to have as few sizes and types (single hung, or casement, etc.) as possible
- Using expensive materials such as brick or stone only in high-impact areas
- Standardizing bathrooms and kitchens as much as possible between models
- Using high-efficiency heating and cooling equipment
- Using evaporative cooling versus refrigeration in climates that are appropriate
- Providing attic fans to handle cooling in swing seasons

Architect #2 revealed more when he said that in his first step in the process, "I need to build my confidence in the knowledge and professionalism of the client." In other words, he felt the need to challenge or question anything in the client's design program that didn't ring true, based on this architect's experience. Ultimately all five architects interviewed said the same thing, though in their own words. They wanted to understand the motives and the experience level of the person or company they would be working with for the next few years. They wanted to know if their client was a leader or a follower and whether their client wanted a design that set new standards or instead were happy with more of the same as what they already had, which, presumably was "not fun."

Owners or developers' strategies are most important to understand. They can choose to try to appeal to the broadest possible market with truly affordable homes or they can decide to be more discriminating for a particular target market. Either way, their architects must fully understand the strategies, and to some degree question them if they see issues. For instance, a strategy that is based on an attempt to duplicate another developer's success may be doomed to failure if the first developer's business plan is not fully known or understood. It's at these times that a challenge is healthy for the process.

Architect #3 mirrored the comments of the first, but Kephart has worked with this man and knew him to be a master of efficient design techniques. The architect once took the preliminary drawings for a model line of 20 or so homes and solved the following inefficiencies:

1. Somewhere around 50 or 60 roof truss configurations that he reduced to 15
2. More than 100 window sizes and types that he consolidated into fewer than 30
3. Numerous bathroom plans and fixture arrangements that could be reduced by 50 percent
4. Ceiling heights and stud lengths that were all over the map that he reduced to three or four for the complete line of homes

This architect was the first to mention knowing what the competition was planning. **Successful design strategies often stem directly from understanding what the competition is up to and more importantly, what they are *not* doing.** This is where he often saw the opportunity to design a truly successful development in a competitive market.

Architect #4 was a design specialist in that he rarely managed the full range of work through construction but concentrated his efforts on the initial design concepts. With this very different perspective he seemed more free to think of things other than the personality of the client. Some of the elements of design that he brought up as important were

- *Historical context:* Context is the focus of this comment. He said he believes that each home or group of homes or townhouses must in some way respect its surroundings and the efforts of people to create those surroundings.
- *Regionalism of style:* This is similar to context but meshing the choice of historical style with local materials. Today regionalism may have more to do with material choices and their uses than anything else.
- *Beauty from the street:* This is an old design philosophy in single-family home design that previously went way too far, becoming façade design. Today beauty from the street is still important but not at the expense of enjoyment of the entire home design.
- *Synergy of inside and outside:* Frank Lloyd Wright was an early practitioner of the concept of environmental design and merging the outdoors with the indoors. Today's compact communities make this even more important in the development of quality outdoor spaces that are literal expansions of homes.
- *Signature:* This architect thinks that the signature of a development is of utmost importance. That signature is expressed in the details of the community as well as in the homes. To him, this is where architecture resides.

Architect #5 openly admitted that she sometimes has preconceptions before hearing the whole story, but she noted, **"The information we receive at the beginning is never complete."** Undiscovered constraints emerge such as subsurface conditions, topography, the economy, or simply details that have not yet been communicated. To her, it's more important to understand the "sacred cows" of the individuals she will have a relationship with for several years. Do they like or dislike something that trumps all else when present? This information helps her to understand and react to the issues that come her way each day.

Kephart concluded from talking to all five that designers trained in the same skills can be very different even though they all see their forthcoming relationship with another person, the client, as the first order of business. Preconceptions are not the issue at all, but how a high-functioning design team is formed that can really impact design for the better. As soon as designers and builders act in partnership with other people, their preconceptions are impossible to hold onto.

HOME DISTINCTIONS

Second or Vacation Homes

Boomers who may have dreamed of owning that mountain cabin or cottage by the shore for as long as they can remember may feel that now is the time to do it while they are still young and healthy enough to enjoy it. Those boomers should be aware of the responsibilities that come with owning a second home:

- Two homes use twice the quantity of natural resources to build.
- Two homes could consume twice the fuel and electricity to heat and cool even when one is vacant.
- A vacant home in the community for long periods of the year can be a burden on community services.
- The location and design of second-home communities should contribute to the enjoyment of that special place by as many people as possible.

Research from the National Association of Realtors in 2006 found that 11 percent of boomers were somewhat to very likely to purchase a new home within the next 12 months. Of that 11 percent:

- Sixty-six percent planned to purchase a primary residence.
- Fifteen percent planned to purchase a vacation home.
- Forty-seven percent of those likely to buy a second home planned to use it for vacation.
- Thirteen percent saw that second home as their primary residence in retirement.

No One Approach

There are many ways to approach building homes that would appeal to second-home shoppers. Many projects marry more than one type of accessibility, for example. Kicking Horse condominiums addressed an affordable market in the North Park Valley of the Rocky Mountains at Sol Vista Ski and Golf Resort. The first two floors of these three-level buildings are at ground level thanks to the steep grade of the site. The third-level flats are accessed by stairs and could be considered better suited to younger buyers, though the 50+ people here are quite active and physically fit. Negotiating one flight of stairs when heading out for a day of skiing or hiking is a small issue. This is especially true with vacation homes where physical activity is often the reason vacationers are there in the first place. Still, accessible homes are necessary for a segment of the population. The lower ground level homes at Kicking Horse provide that necessary accessibility.

Ketchum, Idaho, the city at the base of the ski slopes commonly referred to as Sun Valley, is a wonderful mix of local laid-back charm and a bustling vacation resort. The town separates the Sun Valley resort from the slopes and the Pines development is located at the base of the ski area.

Affordable workforce housing is a huge issue in resort areas, which tend to be on the expensive side, and author Kephart welcomed the opportunity to work on solutions to this problem in the design of the Pines.

The developer/builder was well-intentioned, thinking that these homes at the edge of town would be just the thing to help solve the problem of the lack of affordable housing. Kephart and the planner, David Clinger, presented their high-density clustering concept for approximately 30 homes on this infill site and justified the higher density to achieve affordability, and it was accepted by the community.

Sales started out strong with several homes sold to business owners and town officials, but then market demand from vacationing skiers overwhelmed the process. They offered huge increases above asking prices and outbid local people in competition to buy a home at the Pines.

This story is repeated in resort areas around the country and affordable workforce housing continues to give way to the economic pressures exerted by the visiting tourists wanting a second home in a beautiful location. The Pines became a second-home development by accident, as do many proposals in resort locations. The dilemma for Ketchum and any other resort town is the struggle to provide for their own while continuing to entice tourist trade, the economic lifeblood of the town. Shop-owners and their employees are forced to live miles away and commute to work each day, crowding the small rural roads. Workers in Vail, Colorado, routinely traverse 50 to 100 miles over treacherous mountain passes from their homes in Leadville and other affordable places. Builders electing to enter such markets should be prepared to join with the community in their effort to provide affordable workforce housing.

The Red Quill townhouses in the ski country of Winter Park, Colorado, are three-story homes with provisions for residential elevators (see Fig. 6-5). Few of the 50+ or other buyers have chosen to pay for the elevator option, however. This is a case for installing such elevators as standard.

The developer/builder of Red Quill and the city reached an understanding regarding workforce apartments. For every finished apartment located in the unfinished ground level floors of these three-story townhouses, the city would provide a tax incentive to the developer. In the event that these apartments are not provided, the developer would make a monetary contribution to a fund earmarked to develop future workforce housing units.

Location Factors

An important issue around second homes is where they are located. They are invariably placed in or next to our nation's most beautiful places. Kephart is a backpacker and fly fisherman who loves to lose himself in wilderness areas. He has passed on his appreciation for the solitude and the beauty found in such wild areas to his son and daughter and he is sure his friends are more keenly aware of the threats to these areas because of him.

Figure 6-5
The Red Quill townhouses are three-stories tall with personal elevators offered. (*Photo courtesy of KEPHART.*)

Access to public lands is a huge Western issue because of the large areas of national forests, wilderness areas, national grasslands, and monuments in the Western states. Similar issues crop up around the Florida Everglades, the Great Smoky Mountains in North Carolina and Tennessee, and public lands in every state.

Large, 35-acre lots are permitted in many states without the burden of complying with subdivision regulations, including road building, water, or sewer provisions. For this reason, many land developers have carved up stretches of land along existing roads into 35-acre parcels, blocking access to public lands beyond.

The historical basis for the 35-acre figure as a minimum is spelled out in Verona, Wisconsin's ordinance. Thirty-five acres was considered the smallest piece of land that could support one family by subsistence farming. Regardless of whether farming in Wisconsin is a bit more productive than farming in Wyoming, both states have a 35-acre residential lot exemption from subdivision regulation. Verona's ordinance goes on to say that they do not intend to encourage the division of farmlands into 35-acre home sites, only to allow the continued farming of that land.

Access to public lands is simple to provide with good planning. An easement for public access between two lots, including space for parking and trailhead information will address this issue in part.

Clustering the same number of lots into a smaller area will open public lands visually for the benefit of everyone. Access through the undeveloped part of the private land can still be limited to easements defined with signage. The result is a more beautiful area for those living there and visitors as well. Of course, the development will then be required to go through the process of approvals by the state, county, or other jurisdiction, exactly what these developers are trying to circumvent. It's better planning but as long as these antiquated laws are on the books and people continue to purchase these properties, nothing will change.

Homebuyers, builders, planners, and architects can influence change by refusing to buy these land-wasting 35-acre lots or to build or design homes on such lots.

Regarding the development of rural villages versus privatizing so much of our beautiful forests, farms, or ranchlands, small rural villages are the ideal locations for three out of five boomers in retirement, per the National Association of Realtors. These villages can attract small service businesses looking for a location along a stretch of rural road. The more services that locate here, the richer the lifestyle, and the scenic wild environment is still only steps away. A village can provide a common open space and/or building for gatherings and a quiet social atmosphere for a full retirement life.

Protecting Possessions

To address the issue of how residents keep their possessions safe while renting out their second or vacation homes, "lock-out" units or owner areas provide a secure space for the owners' personal items while others use the rest of the home.

In the case of larger condominiums, design often allows them to be split into two rental units to offer flexibility in pricing for rentals. Renters can take half of the home if that's all they need or the entire space for larger family gatherings, and the like. The local jurisdiction will regulate this type of unit division and may require additional parking and other features to accommodate more people in the development.

Second or Vacation Home Similarities

Second homes share little with each other except for their informality and a few general characteristics:

- Notably missing are formal rooms such as living rooms or dining rooms. The functions that took place in those rooms happen instead in a large great room.
- In active vacation spots such as ski resorts or beach communities, a large vestibule is probably a good idea, to be used for the storage of outdoor gear and outerwear. This vestibule may allow for the showering off of sand from the beach or the removal of snow or mud from boots.
- Kitchens may be minimized since complicated cooking is a waste of time that could be used to enjoy the outdoors, but this cannot be generalized for everyone.
- Bedrooms may include bunkrooms for two to four people each in some cases.

Reverse Second Homes

After retirement and the move from the old neighborhood to their "vacation home" on a permanent basis, people may find that they would like to maintain a small place back in their home city. This reverse second home can be used for weekend visits to family and friends, to take part in social or cultural events, or to continue to conduct business with old contacts.

A small condominium or pied-á-terre is ideal. The small space is easy to care for, all other maintenance is handled by others, and owners can lock and leave with confidence that everything is secure. These homes can also be provided with "lock-away" areas for owner storage, and the home can be rented when the owners are away in their vacation home.

Duplexes

These homes are known by various names depending on the part of the country they are located in; *ramblers* in Minnesota, *villas* in Florida, doubles, paired homes, carriage homes when stacked one above the other, and just plain duplexes. Duplexes minimally increase the density of a development without losing the single-family scale of the buildings, which is the usual reason for using them. In some cases architects and builders try very hard to conceal their true identity by avoiding an obvious pairing of two identical living units. Garages are split and concealed around the corner; entrances are also concealed from any one vantage point, and materials may be changed in ways to make the building of two units still look like a single-family home.

The design of the individual homes, though, very much follows the patterns of designing single-family home plans: they are single story except for perhaps secondary bedrooms, the bedrooms are split, and great rooms are the predominant type of living space.

Triplexes

Less common than duplexes and yet not quite townhouses or row houses are three-unit manor homes. The three-living-unit buildings are treated by building codes the same as houses in many cases and the federal laws regulating accessibility do not apply. Sometimes these manor homes can be used as a part of a strategy when dealing with homeowners or groups of neighbors during the approval phase of a development. When done carefully the three-unit buildings can look very much like large single family homes concealing the small size of the individual homes and the impression of multifamily buildings.

Townhouses or Row Houses

Relatively narrow living units are attached side by side typically in four to eight living units and even larger combinations usually with straight party walls between. The front door and one room usually face the street with the garage, or the garage may be in the rear much like houses in TND communities. The traditional townhouses, such as the brownstones in East Coast cities such as Washington, D.C., Boston, and New York were multiple stories tall to increase the number of living units possible on a small area of very valuable land. Townhouses for the age 50+ homebuyer are created with much the same design intent today but the townhouses in resort-styled active adult communities are mostly one story in height just like the houses in those communities. More multiple-level townhouses are cropping up in urban locations and the inclusion of elevators is becoming more common in those areas of higher land values.

Townhouses provide buyers with a choice in lifestyle different from single-family homes and duplexes. Though the common maintenance provided by the community is the same for single-family homes and townhouses, buyers often have the perception that the townhouses provide a more maintenance-free lifestyle. What townhouses normally do offer is a home at a lower price point than the single-family homes in a neighborhood, giving buyers a way to buy into a higher-value community than they could afford if single-family homes were the only choice.

> Author Kephart discovered an interesting phenomenon about homeowner or homebuyer perceptions several years ago. When attempting to develop a small infill property in a south suburban Chicago community, he and his client chose to design a very compact single-family development in order to avoid the negative perception that he thought townhouses would create in the minds of the people involved in the public hearings. They were 100 percent wrong. The tiny single-family lots necessary to achieve the desired density without proposing townhouses totally confused the citizens. They had a good understanding of how townhouses work but were unfamiliar with the little yards and patio areas for the proposed houses and therefore something to fear. The plan was still successful but the proposal passed by only one vote.

Floor planning for 50+ townhomes again follows the pattern for single-family homes with differences necessary due to the narrow and deep floor-plan configurations.

With a front-loaded garage configuration, as in typical suburban locations, there is little room for more than the front door at street level with units that are 30 feet wide. The width needs to be in the 40-foot range to accommodate a two-car garage, an entry, and a living room or den. This then forces all other rooms to the rear. The great room and kitchen can share the light from 12 feet of the rear wall, the master another 12 feet, and a second bedroom the remainder. In other words, to design a one-story plan with more than two bedrooms and the natural light that boomers want, the plan needs to be even wider, nearly negating the reason to build townhouses. Of course end units can accommodate more rooms since they have more walls and if the program demands more of these they may be accommodated by building more three- and four-unit buildings instead of six or eight living units per building. Building two stories with elevators can overcome these limitations even if you only put a couple of secondary bedrooms up on that level. When the master suite is also on the second level it may be possible to reduce the unit width even more and increase the unit yield on the property, which is the original goal of using townhouses.

In the case with rear-loaded garages you have the same restrictions unless the garage is detached or partially detached to create outdoor space in the form of a courtyard between the house and garage. You start with 30 feet of available exterior wall space for windows on the front, and it's possible to develop three-bedroom living units on one floor in the center of a long townhouse building. Put the secondary bedrooms on the second level and you can flood the home with natural light.

Two-Story Active Adult Homes

Active adult communities are becoming more appealing to buyers age 60 and younger—boomers! From 2001 to 2007 the percentage of households under age 60 increased from 11 to more than 20 percent, according to a 2009 National Association of Home Builders/MetLife Mature Marketing Institute study. Active adult communities generally offer a variety of models, usually including some form of both one- and two-story homes. A huge majority of respondents to several studies of people age 50+, though not specifically boomers, said they preferred a one-story home. With the aging of the boomer population, that trend is sure to continue and increase.

However, when faced with balancing cost and availability over preference, many buyers must choose alternative home types that consume less land and materials and are, therefore, more economical. Two-story homes on smaller lots, for instance, can still have the same yard sizes and will cost less than the same-sized one-story home. Still, the preference for single-story houses is strong. When author Schriener lived in New Jersey a handful of years ago and was in the mood to look for a house to buy, her Realtor told her that one-story ranch homes were so popular that they sold far faster and for closer to their asking price than multistory homes.

The relative affordability of two-story versus one-story homes push many home shoppers over the edge and into a two-story home even if they really prefer a single-story home. Table 6-2 shows why two-story homes cost less.

One-Story versus Two-Story Homes

Try the following exercise using your own figures to test the applicability to your market area. Land costs will be the greatest variable. The large lots required for a 2,200-square foot ranch will yield only a three-homes-per-acre density while the smaller two-story lots will yield 4.5 to 5 homes per acre. Parks and open spaces are not part of these density figures. The figures mentioned in Table 6-2 were made with the assumption that a 25 percent increase in density would be reflected in a relative decrease in lot cost.

For many people, cost is not the greatest issue when comparing one- and two-story homes. Aside from the obvious absence of stairs leading to bedrooms above, single-story homes have other appealing qualities. Access to outdoor patios or courtyards can be given from any space people may want. Patios off of master bedrooms and baths are commonly offered in active adult homes. A single-story plan can feel more spacious than a two-story, since no space is lost to stairs and second-floor corridors. To many

Cost for One-Story Homes	Cost for Two-Story Homes	Table 6-2
One-story home—2,200 ft^2	Two-story home—2,200 ft^2	Example of Costs for One-Story versus Two-Story Home
$80/ft^2 hard cost = $176,000	$70/ft^2 hard cost = $154,000	
Larger lot cost = $80,000	Smaller lot cost = $60,000	
Margin for sales/marketing, profit, overhead = 30%	Margin for sales/marketing, profit, overhead = 30%	
Sales price = $330,000 ($150/ft^2)	Sales price = $274,000 ($125/ft^2)	
Result: the two-story home cost $54,000 less than the same size one-story home.		

Example is for a home in an outlying suburb in metropolitan Denver in 2007.

buyers, though, it may simply be the commonly held belief that a one-story home is more appropriate for them that convinces them.

Qualities of single-story homes can be incorporated into two-story homes, townhouses, or condominiums as well. Fewer rooms in the same home size are a distinctly popular quality in home designs for older buyers. The combination of living and family room into a single "great room" living area is another highly valued feature. Bedroom count is usually lower in active adult housing than for family homes, with three bedrooms being the overwhelming preference. One of those bedrooms usually serves as a den or office and the third bedroom is usually a guest room.

Most newly constructed two-story designs offered on the market today locate the master bedroom suite on the ground floor. The percentage of homeowners in active adult communities who climb stairs to get to their bedrooms in ProMatura most recent study was just 5 percent versus 22 percent of age 55+ homeowners who lived in intergenerational communities. Putting the master bedroom on the first floor gives the age 50+ buyer 75 percent of their preference for a single-story lifestyle. While these "main floor master" homes (MFMs) can be developed on somewhat smaller lots and can cost less to build than a ranch if designed with care, they are only halfway to true affordability. Two-story homes that have the master suite upstairs with the secondary bedrooms need significantly smaller lots and are the most affordable to build. Small lots are not an obstacle to these buyers.

The typical MFM 2,200-square-foot home has only one or two bedrooms and a small loft upstairs. The floor area of the second floor seldom exceeds 700 square feet, leaving the bulk of the space on the first floor, doing little to help reduce lot size and cost. A comparable ranch, complete with a two-car garage, porches, and patios covers nearly 3,000 square feet. The MFM only reduces this number by 700 square feet. In contrast, the full two-story home with all bedrooms upstairs covers only 1,800 square feet or nearly half of the area covered by the one-story home.

Elevators inside two-story homes provide convenience and easier navigation than having to rely solely on stairs. They can make the difference between home shoppers buying a two-story home that's available or continuing to shop for a one-story home (see Fig. 6-6 and 6-7). See more details on elevators later in the section "Elevators".

Accessory Dwelling Units

A small backyard cottage or carriage house above a garage with a personal elevator may provide just the right place at the right price for young adults or older parents to live independently but close by. These cottages or carriage houses are called *accessory dwelling units* (ADUs).

Older homes may have the room for three generations of one family, but accessibility is a major problem for those over 65. One kitchen and one or two baths shared by everyone will, at best, be inconvenient, and at worst, cause ongoing conflicts within the family. Many boomers, the first of whom are now solidly over 60 themselves, still

Figure 6-6

A two-story home without an elevator. (*Drawing courtesy of KEPHART.*)

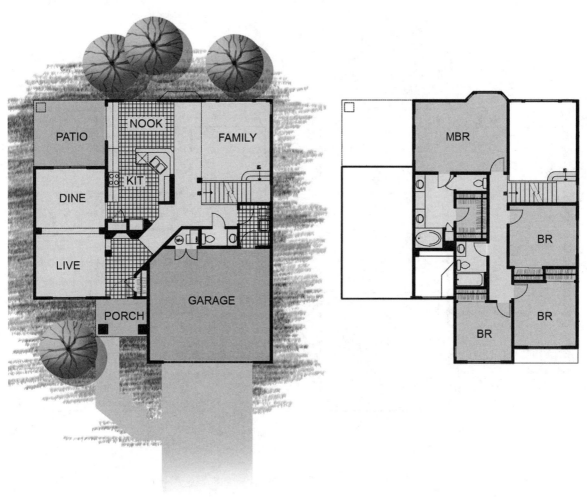

have adult children living at home not yet able to afford their own place. They may also have aging parents beginning to need care and a safer housing environment. Housing in an independent living community can cost $3,000 to $6,000 per month. Costs for assisted living are even higher. This expense puts the boomer families in a crunch between supporting their children and their parents, earning them the nickname "the sandwich generation."

Some families are joining Mom and Dad in the larger family home, while others are asking the kids to move to the basement to make room for parents and grandparents on the first or second floors. The physical problems with these older homes include the difficulties that stairs, small doors, and narrow hallways present to older adults, especially those who require the assistance of a walker, wheelchair, or other device.

Figure 6-7

Same plan as Fig 6-6 with an elevator to attract the 50+ buyers. (*Drawing courtesy of KEPHART.*)

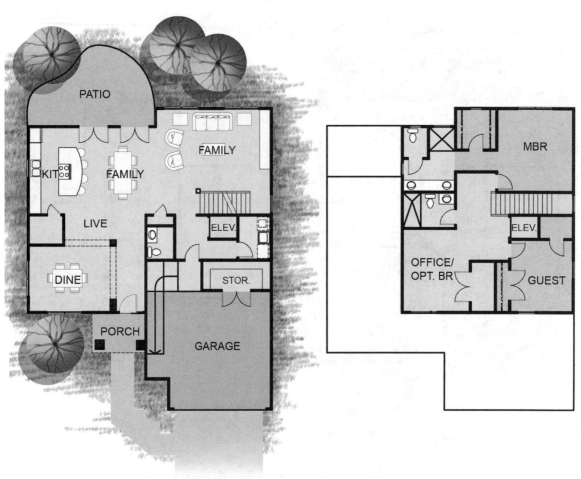

According to the latest U.S. census, the number of households with three or more generations living under one roof grew 38 percent from 1990 to 2000 versus 8 percent for households containing just two generations. The growth rate of this market has shot up even higher, as much as 60 percent, since 2000, and with the aging of the boomers it is sure to absolutely skyrocket over the next decades.

A carriage-house-style ADU behind a single-family home can provide a place for an aging parent to live independently but still close by in case they need assistance. Or, all the comforts of a home can be packed into a 700-square feet home above a garage (see Fig. 6-8). An elevator is a modern addition that can make this home accessible

Figure 6-8

An accessory dwelling unit is often located above a garage in neighborhoods with alleys. An elevator may be required for those who have difficulty with stairs. (*Photo by Mike Kephart.*)

to people of all ages and abilities. For older citizens living alone, it could provide a modest living space for a caregiver to help them if family members are not available. Single women and men facing forced retirement in their early 60s could build a carriage house for their residence, allowing them to lease the primary house to a young family that cannot yet afford to buy a home in the city. People with older married children that are unable to make it on their own can provide a temporary refuge at a nominal cost compared to buying a home for them. ADUs can be rented when a parent(s) needs to move out of the home setting to receive more care than the family can provide. This rent may cover the payments for the ADU, plus provide additional income to help provide care for parents elsewhere.

There are countless stories from people who need a way to improve their property, provide living space for a disabled child or relative, or to simply help out sick relatives who need a place to stay until they get well.

An ADU is known by many names: granny flat, carriage house, an in-law suite, and more. All the various names are intended to characterize a second living unit complete with kitchen and bath. Most communities require that such a living unit be accessory to the primary single-family use or rented by the homeowner. It may be included as part of the primary structure or detached. It may be located above a garage, on the rear of a single-family lot, or located anywhere on the lot that meets local zoning regulations. ADUs are usually small, anywhere from a 400-square-foot studio unit to a one- or two-bedroom unit with 1,000 square feet. Regulations vary on size, number of bedrooms, lot size needed, and design (see Fig. 6-9).

Figure 6-9

A 600-ft² floor plan of a one-story accessory dwelling unit. (*Drawing by Mike Kephart.*)

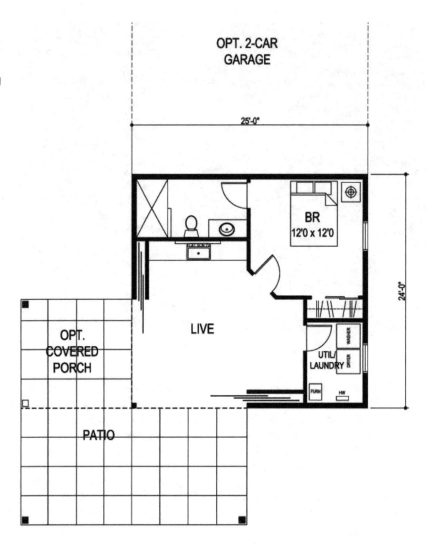

FLOOR PLAN
600 s.f.

Zoning Challenges

The inclusion of carriage homes in zoning ordinances would diminish the pressure to alter the form of existing homes and neighborhoods should a property owner choose to add living space for the family. It's also a way to bridge the gap between the higher-density zones and the low-density single-family areas. Homes near light rail stops and other transportation hubs could accommodate a small addition to the population in that area without anyone having to demolish existing homes. The streetscape will not be impacted by this type of addition as opposed to "pop tops" or second floor additions that have been so common in the past.

Carmen and Uncle Paul

The complete story of the need for ADUs can't be told without meeting the people who are facing these changes in their own families. The following is just one of those stories.

Carmen and Ed lived in one of Denver's venerable old neighborhoods. They shared their small two-story Victorian home with their 3-year-old daughter and were expecting their second child. Ed worked in the technology industry and Carmen was a stay-at-home mom.

Last year, they invited Carmen's 62-year-old uncle, Paul, to move from Arizona and live with them now that he was retired from his job of 16 years as a computer systems manager. Earlier, when Ed accepted relocation to Europe from the same company, Uncle Paul took the company's offer of early retirement.

Carmen had reasoned that adding a living space above their existing brick garage would be a simple way to make room for Uncle Paul. This would keep him close by while giving him his independence, although he didn't handle stairs too well. They, like many before them, soon discovered that Denver's zoning regulations did not permit the building of such an addition to their property.

Jack wanted an ADU for his mentally challenged adult son to live on his own, but his proposal was rejected. Carmen wanted to house her aging uncle in a carriage house she planned above her garage, but she also discovered that the existing carriage houses, common in Denver's older neighborhoods, are no longer permitted. Bob's goal was to block sounds from neighborhood restaurants and bars across the alley from his backyard by building a garage to include a second-floor rental unit. The rent would also help him pay for his retirement and allow him to remain in the neighborhood where he had lived for years, but was growing less and less affordable. Stories like these are repeated in increasing numbers. Antiquated zoning ordinances that date back nearly 50 years are clearly not keeping up with family composition changes or cost of living increases for families today. The administrators of the code were responding to issues of that earlier time, but that is no solace to a family in need of relief now.

Santa Cruz, California, started a formal revival of the granny flat in 2002 with its new accessory dwelling unit zoning ordinance. Essentially it removed the barriers set up from a time when people were more worried about overcrowding than how to encourage people to move back into the close-in residential areas surrounding city downtowns. Times have changed, and Santa Cruz officials saw this move as one way to help families come together to nourish their aging citizens, to provide affordable housing and to bring back the population lost to their city during the great flight to the suburbs.

Last year author Kephart began working with an ad hoc neighborhood group of Denver residents that had all run afoul of local zoning regulations while trying to provide housing for their extended families. The group now is known as, "Friends of Granny Flats." A resident in Kephart's neighborhood called the group together at a coffee shop where they meet at 7:00 a.m. every Tuesday morning. Each member had attempted, in their own way, to set up a second living space for an aging parent or other relative in their single-family home. Several people simply wanted a parent or parents to move in with them and be able to live independently with their own kitchen, bath, bedroom, and living areas. However, separate living spaces constitute a zoning violation in Denver's single-family zone districts. Fortunately, Denver and other cities are changing and writing new, more favorable regulations that will permit ADUs.

Currently hundreds of cities have revised their rules to allow small accessory dwellings as a second home on a lot, providing that the addition meets certain regulations. The entire state of California has followed the path set by Santa Cruz, and Oregon, Virginia, and Washington state are in the process. Other states are following suit city by city and county by county. Colorado, for example, has a minimum of 12 cities and counties that have adopted similar regulations and more join in each day. Other cities that now allow ADUs are Austin, Texas; Pensacola, Florida; Fort Collins, Colorado; Sedona, Arizona; Casper, Wyoming; Portland, Oregon; Rindge, New Hampshire; Minneapolis, Minnesota; and San Antonio, Texas. They see this movement as one solution to the affordable housing crisis and to housing our rapidly aging population. See the Appendix for the regulations for Portland's zoning for ADUs. Most city regulations are similar to Portland's but variations exist in all of them.

The U.S. Department of Housing and Urban Development (HUD) published a case study on ADUs in June 2008. You can find this case study at http://www.huduser.org.

INDIVIDUAL HOME FEATURES

Breaking down common elements of all types of homes will provide you with the benefit of the experience of those familiar with what age 50+ homebuyers prefer and how builders can incorporate those desirable elements into the homes. Room-by-room design is covered in Chap. 8.

Ceiling Heights

Builders and architects have experimented with ceiling heights, vaulting, and dramatic shapes and relationships between floors for homes designed for the 50+ market, but now that they have made all of those attempts, most have settled on simple flat ceilings of 9 to 12 feet high depending on the room sizes and home sizes. Many boomers see vaulted ceilings, shelves, and niches above easy reach with a small stepladder as having pesky maintenance and cleaning issues. Home shoppers typically resist builders' temptation to connect small second-floor areas with the first floor of a 50+ home through dramatic ceiling shapes because of its complexity. The homes with the master bedroom on the first floor are designed for older buyers who perceive secondary rooms on second floors as simply utilitarian rooms for visitors or grandchildren. They don't see these rooms as part of their daily living space.

Views

- Homes in vacation resort–styled communities are typically designed primarily around views to private rear yards, patios, and open spaces beyond.

- Homes in TNDs are primarily designed to emphasize the street presentation of the elevation. Views are to more intimate courtyards or patios in side yards. The individual home design is less important than the community appearance as the homes are seen from the street.

- In designing conventional suburban developments, architects work very hard to reduce the negative impact of the front-loaded garage on the overall impact of the home design. Garages are pushed back behind the front face of homes to diminish their presence. Sometimes side-loading a few garages in a block can add variety even though it adds paving for a larger driveway at these houses.

In those areas of a master planned resort-styled community where homes must be placed back to back with no view opportunity except for the back of a neighboring home, a few special floor plans can be offered that do not rely on rear views. Courtyard homes that contain open space inside the footprint of the home will direct all views to those courtyards, creating particularly private outdoor space. When designing courtyards it is important to design the elevations of floors inside and out all at the same height. This not only avoids the use of steps, but acts to blur the lines between indoors and outdoors. Then think of that courtyard as another room and encourage the homebuyers to furnish it completely as they would any room, even hanging paintings on the walls and putting throw rugs on the floor. This attention to detail will make the entire home feel larger.

Elevators

Elevators, once unheard of in home design, are growing in popularity. Researcher Margaret Wylde presented research findings that were new in her presentation at the 2007 International Builders' Show. She reported that 56 percent of boomers would select a second-floor master bedroom if an elevator were provided in the home. Elevators are commonplace in high-end homes in infill locations today. At approximately $20,000 to $28,000 installed, depending on options and finishes chosen, elevators deliver prestige, of course, but also comfort, convenience, and, surprisingly, affordability. Elevator manufacturers are offering quality residential elevators with spacious cabs and smooth, quiet equipment and doors.

Building on the example earlier of home costs for comparable one-story versus two-story homes (Table 6-2), assuming no elevator in the two-story home, Table 6-3 is a continuation of calculations with an elevator factored in.

Active Adult Homes Construction Costs

Two-story homes that cost $26,000 less than a ranch and have a convenient, well-designed floor plan including an elevator should catch the attention of 50+ buyers.

Table 6-3

Hard/Construction
Costs for a 2,200-
Square-Foot Home

	Ranch	MFM	Two-Story
Without elevator	$176,000	$165,000	$154,000
With elevator	–	–	$174,000
Density	4 homes per acre	5 homes per acre	6 homes per acre
Sales price	$330,000	$312,000	$304,000 (with elevator)

Note: Above figures assume conventional lot layouts, two-car garages, and front and rear yards. No clustering or higher-density ideas were included. Figures are for illustration purposes so verify in your local market.

If anything, the figures detail in Table 6-3 are very conservative. The point is that two-story active adult homes can be cost-effective.

As for elevators in home design, consider two concepts. You can feature the elevator as part of the upfront entry statement or you can locate it more subtly and conveniently near the kitchen and family room. Regardless of placement, the authors strongly recommend installing an elevator as a *standard* item in the home. Offering it as an option aggravates the buyer at the idea of having to pay extra for it, and not having it could drive the buyer away to look elsewhere or create regret after moving into the home.

Builders have used elevators as options in three-story townhomes, but the fact that they were not modeled in the sales park as well as being an extra-cost item gave buyers no experience of how easy and fun living in a home with an elevator can be. Pay a little extra to upgrade the cab size as well. It's a small extra cost that will pay off in sales.

Don't be your own competition! If you mix ranches and two-story elevator homes throughout the community, you will lose much of the savings in land cost by way of more compact development. At least separate higher-end ranches from the more affordable and compact two-story homes so smaller lots can be planned for the two-story sections. It's the land and the construction costs together that make this concept work.

Active Adult Two-Story Elevator Homes Dos and Don'ts

DO: Choose a quality elevator that is smooth, quiet, roomy, simple to operate, and beautiful.

DON'T: Choose a cheaper, smaller, slower, noisy machine. Buyers' doubts will be confirmed.

DO: Include the elevator as standard equipment and show it in your model so people can use it.

DON'T: Offer the elevator as an option. Buyers will focus on the cost and not the benefits.

DO: Create at least a small two-story space around the stair and the elevator to show off both floors at once and make it easy for buyers to understand.

DON'T: Put the stair up to the second floor right at the entry. Let the buyers experience the elevator first, before they see that the second floor is there.

DO: Design the front elevations of homes to appear to be one-story as much as possible. You need for buyers to come in to get the story.

DON'T: Compete with yourself by offering both ranches and two-story homes in the same location. You'll be tempted to increase all lot sizes and lose the affordability.

Personal Touches

Allowing buyers to design something special for themselves in their new home will create a source of pride for them for years to come. The task is to find a way that you can work it into your system of design and building. For many it may be as simple as the pattern of tiles above the kitchen counter or a small niche for special heirlooms. People need to be able to design their own special feature or two. Options alone won't do, because the point is to allow *personal* expression.

Homebuilders in smaller-volume market areas, such as Cleveland, Pittsburgh, Cincinnati, and many Midwestern and Southern cities are far more willing to make significant changes to their home designs to please the fewer buyers they attract in a year. In Cleveland, an area with which author Kephart is very familiar, a volume of 30 homes per year in a development is quite common. Numbers like that would be considered a failure in California or other Western or coastal marketplaces.

QUALITY TRIUMPHS OVER QUANTITY—FINALLY!

The boomers may finally begin the longed-for transformation of American housing from a per-square-foot commodity into places to live that are valued for their quality, durability, sustainability, and ability to comfort and nourish the soul.

Ken Dychtwald, the gerontology guru and demographer, reported in his book *Age Power* the results of a poll conducted by Roper/Age Wave (Age Wave is Dychtwald's company) in which the new boomer American dream included top priorities of "being true to self, not selling out, and achieving inner satisfaction" versus power, influence, or wealth, values that were all at the bottom of their priority list. Ironic, isn't it, that the priorities boomers now eschew are the very ones that they brought into favor and around which they created the workplace of the last four decades. Whether it's dissatisfaction with where that got them, wisdom gained from experience, or just a natural evolution tied to getting older, homebuilders need to pay attention to boomers' new priorities.

Some ways that these priorities manifest that are relevant to homebuilders are

- Boomers are not just finding more time to enjoy life. They are actually participating in life more deliberately, more fully, and with more presence of mind.
- They feel less need to impress or to define their status with their home. They have more honesty, humanity, spirituality, and humility. Their home, no longer merely a transitory investment property, can finally express their true spirit and values.
- Clearly, they need less space since the family has grown and moved on. (The exceptions are those whose parents or adult children share their home—more about them later in this chapter.) Moreover, they desire a smaller home built with quality and care. Boomers want to simplify their lives, live informally, and spend more time with friends and family as well as build new relationships with like-minded people.
- They now have time to put themselves into the design of the home they will live in for the new adventure of the next phase of their life.

The Scourge of the Big Box

The authors hope and are planning that the priorities mentioned earlier will result in the demise of the notion that the "big box" is the way to go. No concept in home design has retarded good design more than the idea of the big box. Here's how it works:

First, the home must be two or more stories tall since a two-story house will cost anywhere from $10 to $20 per square foot less to build than a one-story house of the same size.

The next principle is to size the first and second floors to fit precisely on top of one another to eliminate building corners, a known costly element though difficult to quantify. The problem is that the square-foot area needed for the first and second floors is nearly always different. One or the other must be increased in size to make this idea work, since reducing the size of any room would no longer meet the design parameters.

The result is that anywhere from 10 to 20 percent more space is added that wasn't needed to satisfy the design program. Rooms are needlessly enlarged, circulation space grows, and unnecessary spaces such as "bonus rooms" are added.

The house does cost less per square foot but the finished product is too big. Quality is sacrificed for size and then size is sold to the buyer as "value." The home is often an unadorned "box" with all of the charm that the word brings to mind.

This is not to say that simplicity doesn't have a beauty and value of its own. It's just that the "big box" idea has been pushed to the extreme. Ideally, simplicity in design saves the unnecessary costs ascribed to complex plans and uses those savings for higher quality, more durability, and maintenance-free materials and details—just the things boomers say they want.

On the other hand, extra corners in the smaller, not-so-big-box plans create the opportunity for outdoor rooms intimately connected with other rooms. The more complex shape of the plan also breaks the "box" into more human-scaled architectural forms.

An Affordable Alternative

Prefabrication or modular building is an affordable construction idea that is far more flexible and effective. Consider these advantages:

1. Aside from lot preparation, excavation, and foundation construction, everything else that goes into a home can be built inside a factory in a controlled environment.

2. Material usage is more efficient, quality control can be assured, and the workplace is safe and healthy for the workers. Architectural designer Rocio Romero has a small factory in a rural town in Missouri where the labor costs are low and new jobs are a welcome. Her precut homes come complete with all of the materials, equipment, and finishes needed to build on a foundation. The entire home down to the faucets and windows comes to the jobsite on one truck. Assembly instructions facilitate the completion of the home in a very short time.

3. Other companies build complete sections of homes, transport them, and assemble the modules onsite, and others "panelize" walls, roofs, floors, and other building components. All of these methods are designed to minimize the time construction crews are needed onsite, saving both time and money.

4. Once complete factory-built homes have been delivered to the site, little work is needed beyond anchoring the home on the foundation and connecting wiring and plumbing.

Opting for Quality

More than half of the respondents in Wylde's recent survey of age 45+ middle-income Americans who were planning to move in the next 5 years said they would choose quality over size, that they preferred a smaller, high-quality home over a larger home. Those 55+ who planned to move to an active adult community felt even more strongly about quality. More than two-thirds said they preferred a smaller, top-quality home to a larger home. The prices of homes also made a difference. Nearly half of respondents who were willing to spend less than $150,000 on their home voiced a desire for as much space as their money could buy. On the other hand, a little over one-fourth of respondents who were willing to spend more than $400,000 wanted as much space as they could get, the rest opting instead for a higher-quality house. Home designer Phelps said one of the advantages his clients see in buying a smaller home is that they can get higher-quality materials and fixtures for less than they would have to spend for the same things in a larger home: "They're able to get high-end luxury items for a smaller square footage. They love it."

Smaller Homes

In one of Wylde's studies, 22 percent of boomers said they wanted a home sized 1,000 to 1,499 square feet, 32 percent said they preferred 1,500 to 1,999 square feet, and 21 percent preferred 2,000 to 2,499 square feet. Only 22 percent wanted a home larger than 2,500 square feet. It all adds up to providing smaller homes for the 75 percent of 55+ homebuyers that say they would prefer a smaller house with higher-quality products and amenities.

Sarah Susanka in her book *The Not So Big House* (see Fig. 6-10) and subsequent publications struck a resonating chord with homebuyers. The essence of her "not so big" philosophy is not necessarily the total size of a home but a home's size in proportion to the lives of those living there.

Most 50+ boomer couples will continue to work, though at a reduced pace. To facilitate this new lifestyle, each partner may want their own workspace or hide-away—one of

Figure 6-10

The Not So Big House caught the imagination of the public and justified building smaller and greener for thousands of Americans. (*Photo courtesy of Sarah Susanka.*)

Susanka's ideas—for their own good as well as for the good of the marriage or partnership. Visiting grandchildren are also uppermost in the minds of boomers. The home should accommodate places for privacy and work, visiting grandchildren, entertainment, and the couple's own shared bedroom suite. This could mean more rooms and a larger house unless you design flexibility into the home for shared and infrequent use of the same spaces. Following are some ideas as to how to do that:

- Basements or attics can be great places to house visitors and grandchildren but there are ways to include everything on the single level, which survey respondents overwhelmingly preferred.

- A somewhat larger guest suite can accommodate a couple with children or even several children if planned carefully.

- Workspaces and offices for personal use need not be as large as a bedroom and can be partially open to other spaces. Low walls, shutters, moving partitions, or sliding doors can conceal work clutter.

- A bunkroom can happily house a group of youngsters without consuming enormous amounts of space.

- Flexible uses for these areas as hobby rooms or simply as part of the living or family areas can reduce the need for additional, seldom used space.

Depending on the type of community the home is in, there are uses that once required space in the home but are now provided either in the community club house or close by in the community. Fitness and weight-training equipment, for instance, is state-of-the-art in most community centers and a trainer may be employed for instruction in the use of machines. Some communities even have guest rental apartments available for visiting relatives and friends. Large-scale entertaining can take place in a reserved space in the clubhouse or in a community garden instead of in the home. Parking is more plentiful and convenient at the community buildings and the community may offer catering services.

The challenge for the housing industry is how to take Susanka's "not so big" principles and tailor them to fit production housing business practices as well as to develop ways to incorporate more personal attention into each home built. A generic model 1, model 2, or model 3 with limited options cannot adequately involve the buyers in the process. Remember the values mentioned earlier: quality, durability, sustainability, and comfort and nourishment of the soul.

Rightsizing

Ciji Ware's book, *Rightsizing Your Life,* is a consumer workbook on how to simplify life for those who suddenly realized that life is running them rather than the other way around.

Just as in Susanka's "not so big" principles, the issue is not how big a home is, but how to size it in proportion to a family's needs and to avoid the wasteful use of space, material, and energy. There is certainly nothing wrong with a couple's decision to continue living in the home where they raised their family, even if the house is really too big for them now that the kids have left. Family memories reside there and every room has a story to tell of the trials and joys of raising a family. However, a couple building or buying a new home is different. They can look closely at their lives and make the decision to "rightsize" their life, as Ware put it.

Deciding whether a home is too big or too small is a subjective thing best left to informed personal judgment. As builders and architects, you can help educate people who are open to discussing these notions. You can show buyers alternative ways to accommodate their perceived "need" for additional or extra-large spaces. You can help people understand and appreciate the value of design qualities such as intimate, cozy, warm, and homey, all feelings that are difficult to create in huge, barn-like family rooms. You will notice that the techniques for "rightsizing" mirror those of Susanka's "not so big" process as far as the house is concerned. Ware goes on to outline all parts of a couple's life together and many things, such as clearing out unneeded furniture, and long-forgotten stored items, come back to impact the planning of the home.

Cynthia Dickerson, former Director of Marketing Research for John Laing Homes in Denver, Colorado, has given a lot of thought to downsizing and rightsizing. According to her, the common industry wisdom that 50+ buyers are generally looking to "downsize" when buying a new home didn't match buyer comments and actions in her experience. While 50+ buyers may be downsizing their life by downsizing the home maintenance needed, the household headcount, or the financial outlay, they are not necessarily reducing their home size, she says. Recent research by the National Association of Home Builders supports her perceptions. Paul Emrath, NAHB's lead researcher, said, "Our data shows that 55+ home buyers may be 'downsizing,' but not by much. The average home in an active adult community still includes more than two bedrooms and more than 2,000 square feet of living space." Understanding this mindset in prospects has helped Dickerson's company to better communicate with those potential buyers. "The language use sets the tone for the discussion," said Dickerson. "Rightsizing" has no judgment in the term to imply anything other than what is "right" for the homebuyer.

Dickerson also pointed out that those in the homebuilding business become too set in their ways about labeling rooms as if only one use can be made of a space, such as a bedroom, study, or dining room. One room on a plan was labeled "gift wrapping," to illustrate an extreme example of made-up room names. If they simply referred to space 1 through 10, or whatever, buyers could decide for themselves how to use those spaces. John Laing is planning to test this idea by merchandising spaces in different ways to signal to buyers that any way they want to use a space is "right" for them.

Builders will do well by providing space for the "pain points" in people's lives, no matter how big or small the home may be. A mudroom, for instance, serves as a transition from inside to out and out to in. Coats may be stored there along with boots, umbrellas, car keys, recycling bins, etc.

If a mudroom is not included, the need is still there—it is a "pain point." People will use the back hall, the kitchen, countertops, the stair, and the dining table to scatter these items. Other examples are laundry rooms too small for folding clothes that result in a dining table covered with folded underwear, socks, and towels, and no place for a home computer, leaving no choice other than to put it and its clutter somewhere inconvenient and open to view. Human needs must be met in the most fundamental ways or a house cannot be right sized.

Wasted Space

An overly large family room is probably the most common space waster in the typical home. Maybe the extra space makes sense for people who frequently entertain large groups, but for the majority who rarely entertain on that scale, large parties could take place in a space in the clubhouse designed for such gatherings. For most of us, a more intimate family room, sized "just right" for a group of four to eight, may work best and would be far more comfortable for a couple alone on a day-to-day basis.

REFERENCES

1. Cross, Tracy, Tracy Cross & Associates Inc., Schaumberg, Ill.
2. Del Webb, "Baby Boomer Report," Pulte Homes Inc., Bloomfield Hills, Mich., 2003, 2005.
3. Dychtwald, Ken, *Age Power: How the 21st Century Will Be Ruled by the New Old,* Tarcher/Penguin, New York, 2000.
4. Keister, Lisa, Ohio State University Professor of Sociology.
5. Sage Computing Inc., "Accessory Dwelling Units: Case Study," U.S. Dept. of Housing and Urban Development Office of Policy Development and Research, Reston, Va., June 2008, http://www.huduser.org/portal/publications/PDF/adu.pdf. Accessed December 2009.
6. Susanka, Sarah and Kira Obolensky, *The Not So Big House,* Taunton Press, Newtown, Conn., 2001.
7. U.S. Census Bureau Data, 2000–2009.
8. Ware, Ciji, *Rightsizing Your Life: Simplifying Your Surroundings While Keeping What Matters Most,* Springboard Press/Hachette Book Group USA, New York, NY, January 2007.
9. Wylde, Margaret A., "50+ New Home Buyers: Why, Where, What and When," Research study presentation at International Builders' Show, Las Vegas, Nev., January 21, 2009.
10. Wylde, Margaret, *Boomers on the Horizon: Housing Preferences of the 55+ Market,* BuilderBooks (National Association of Home Builders), Washington, D.C., 2002.
11. Wylde, Margaret A., *Right House, Right Place, Right Time: Community and Lifestyle Preferences of the 45+ Housing Market,* BuilderBooks (National Association of Home Builders), Washington, D.C., 2008.
12. Wylde, Margaret A., "What Boomers Want," Presentation at International Builders' Show, Orlando, Fla., January 2007.

7 Condos and Apartments

"I installed a skylight in my apartment. The people who live above me are furious!"

—Comedian Steven Wright

Not everyone prefers a single-family home, and that includes baby boomers. Boomers' aging is sure to infuse more life into the multifamily scene. The numbers of boomers that will choose apartments or condos probably will climb steeply as more of them become empty nesters. Boomers desire to travel and the idea of being able to just lock-and-leave should appeal to them because it's much easier and safer in a multifamily complex than when they own a single-family house. And, more boomers who are longing to be rid of the responsibility and expense of keeping up a home will be drawn to the so-called carefree living offered by apartments and condos.

At first glance, the prospects for a viable apartment/condo market look a bit discouraging. Over 90 percent of Americans with what can be considered middle incomes between ages 55 and 64 own their homes. The condominium market has steadily plummeted the last few years as the number of investors willing to buy condos and resell them for a profit dwindled. Exacerbating the problem are Fannie/Freddie Mae's newly tightened lending guidelines for condos to limit the lenders' exposure to what they consider a riskier part of the market. That may be good news for the apartment market.

In fact there is always a tight relationship between the condominium and apartment markets during shrinking or growing market tides. In very good times both do well, and in very bad times they both do poorly, but as a good overall housing market begins to shrink, it is condominiums that suffer first and most deeply. The former condo buyers either choose to purchase single-family homes or to rent for awhile. While this reaction further depresses the condominium market, it feeds into the apartment market, usually for 2 or 3 years longer, at which point the downturn will catch up to the apartment market as well. Conversely, as a poor market begins to grow, it is condominiums that continue to lag behind as buyers capitalize on good buys in the single-family market.

Apartments usually do well during this growth transition as those uncertain of the stability of the expanding housing market continue to rent.

Despite the fact that about 61 percent of apartment dwellers over age 55 have an annual income of $30,000 or less, mature market researcher Margaret Wylde sees a market for rentals. One architect who has specialized for decades in multifamily projects encompassing multigenerational, age-targeted, age-restricted, independent and assisted living, affordable housing, and market-rate multifamily projects agrees. Ed Hord, senior principal at Hord Coplan Macht Inc. in Baltimore, Maryland, has seen a trend toward more market-rate active adult projects. As financially secure boomers' kids leave home, leaving them free to downsize to something more compatible with their lifestyle, that trend is certain to continue.

Wylde studied middle-income renters—whom she defined as those age 55 to 64 with annual incomes of $50,000 or more and those 65 and older with an annual household income of $30,000 or more—and estimated the current market for middle-income renters to be over 1.6 million, figuring 7.4 percent of 22 million total households in that demographic. Unlike the heavy concentration of married couples who purchase single-family homes, a majority—62 percent—of renters are single, according to the study. Currently, 64 percent of those renters are women, 36 percent men.

The questions are how much like detached single-family homes do boomers want their condos and apartments to be, what are they willing and not willing to give up to move into them, and what are the key motivators that would have them choose that kind of space over a single-family home.

Answering such questions for your market will give you valuable guidance as to how to design and build spaces that will draw boomers to your project and keep them happy once they are there. Here again the differences between apartments and condominiums begin to show themselves. The buildings for each can literally be identical in appearance, and both must be designed to the lifestyle of their specific market segment, but after this the similarities diminish:

- Apartment renters will settle for a slightly lower level of quality in materials both interior and exterior. Investors in apartments want higher quality so their investment will have a longer lifespan, but they will make compromises in order to compete in a market with older apartment stock.

- More repetition in design is usually a trademark of apartments versus condominiums. With apartments, whole buildings are repeated down to the last detail, fewer apartment model types exist, and few individual design details are to be found.

- Parking in apartment developments usually includes fewer garages or covered spaces and rarely do individual living units have attached garages. Developers of condominiums however, try very hard to provide parking that is secure and out of the weather whenever possible.

- On the other hand, apartment developments often keep more staff on hand for maintenance and repair, while condominiums depend on the individual owners to take care of their own living unit inside. In the most recent experience one author has had with

Figure 7-1
Apartments and condominiums all grouped around a suburban shopping mall on a new light rail line that leads directly to downtown. (*Photo courtesy of KEPHART and Steve Hinds Photography.*)

apartment living, the staff would even come and change the light bulbs when one burned out.

- A central city downtown or a suburban city center can offer a lifestyle of activity and involvement. There you will find a plethora of restaurants, shopping options, entertainment in the forms of movies and performing arts centers, art galleries, and sports venues. More recently, food stores are being added as housing numbers grow. Everything is within walking distance, including hundreds of businesses and offices for those continuing to work. See Fig. 7-1 for example of apartments and condos clustering around transportation.

- Do not expect to find active adult communities wrapped in an urban skin. Most current downtown housing is designed to appeal to a wide range of people: young, old, families, everyone. Suburban city centers tend to offer an "urban-light" mix of housing, including townhouses and, more rarely, single-family homes. To find a building or community with the lifestyle and amenities you want, you'll need to do a detailed search. A Realtor can help get you started but you will need to do some work to find the place for you.

- So many urban-styled developments are built by a group of investors, not builders familiar with the market and the needs of people over 50. Advertisers too often direct ads to a young, hip group when marketing downtown housing, which is a large part of the market, but baby boomers constitute 50 percent of urban homebuyers.

- Be receptive to doing projects that have a choice of home types. Each group of homebuyers wants and needs a unique set of features and designs. A home for a 20-something fresh out of college should be designed differently than a home for a 50+ couple whose children have just left home.

APARTMENTS OR CONDOMINIUMS: WHAT'S THE DIFFERENCE?

Before we go into more on the location-driven aspects of both apartments and condominiums let's look at the similarities and differences between these often-confused housing types. The market for both apartments and condominiums does lead developers to choose locations that would work for both: close to a regional mall and/or on major routes of transportation, highways, rail lines, and bus routes.

The similarities are numerous, especially when the market is for the same income group, and many suburban apartment developments and condominiums are both targeting the entry-level market. Apartment communities tout the freedom renters have to move quickly and the relatively minor commitment people must make as compared to buying a condominium. Condominium marketers emphasize the equity build-up buyers will have and how they will lose that every month they continue to pay rent. This head-to-head competition goes on and on. During good economic times with real estate increasing in value each year, the condominiums may win more buyers, and conversely when economic conditions cause a lack of consumer confidence, then rental looks like a safer bet for people. Apartment developments are even sometimes constructed with the intention of conversion to condominiums when the next boom occurs. That way the developer can rent apartments during slow times and then sell the development to buyers when times are better. That's one reason buildings housing both apartments and condominiums often are impossible to distinguish from each other.

When author Kephart designed his first apartment developments over 30 years ago, things such as having a microwave as standard, or a space for a washer and dryer would separate one development from another. He added other features throughout the years to set his projects apart, including installing individual heating and cooling equipment and individual gas and electric metering; enlarging patios and balconies; adding islands in kitchens, which were a major hit seldom seen in other apartments. Then he moved on to adding living unit choices beyond the old one or two bedroom each with one bath to the split master bedroom units for sharing rent between two people who were relative strangers. The two-bedroom/two-bath living unit attracted couples who each then could have their own private bathroom. The one bedroom plus study even today works particularly well for a single occupant who needs an office.

Today the building designs are of great importance to both buyers and renters. Figure 7-2 is the Riverbend Apartments and Fig. 7-3 is from the Vallagio Condominiums in Denver Colorado, both high-quality, low-maintenance developments.

Apartment owners want buildings that will last for years with a minimum of maintenance, and so do purchasers of condominiums. The competition goes on without pause as developers and architects look for the new feature or the new look that will give them the edge over their competition both from other apartments and from similarly priced and constructed condominiums.

The differences begin to separate apartments and condominiums when the target market is a more affluent group. Yes, there are people who will continue to rent even at a higher

Plans "open" at the rear to take advantage of spectacular views.

Figure 7-2

A high-quality apartment development in upstate New York. (*Photo courtesy of KEPHART.*)

price, but there are more who will choose to purchase when they can afford something of a higher quality of finish that has more space (see Fig. 7-2, 7-3, and 7-4). Transient renters traditionally are waiting until they can afford to buy and will not move up in the rental market unless they are faced with a long time to accomplish their goal. Buying a better,

Figure 7-3

A similar-quality level condominium in Colorado has more durable materials, more personal parking, larger living units, and upscale interiors in comparison. (*Photo courtesy of KEPHART and Steve Hinds Photography.*)

Figure 7-4

Interior of the Vallagio condominiums shows the higher-quality materials and finish more often found in condominiums versus apartments: granite countertops, wood floors, and larger spaces. (*Photo courtesy of KEPHART and Steve Hinds Photography.*)

larger condominium, however, is more often seen as the first-time home purchase or the first move-up for some. Condominium buildings are made of more durable brick and metals, parking is moved underground instead of scattered over the surface between buildings, and some living unit plans offer everything buyers could find in a house or townhouse.

THE NEW WORLD OF CONDOS AND APARTMENTS

The days of home ownership as the ultimate goal for every family may be drawing to a close, or at least diminishing. Boomers are more nomadic than their parents and want more freedom to explore the world or move when an opportunity comes up without having to worry about selling a home. Many boomers are also aware that living in far-flung suburbs with few services available is an increasingly unsustainable way of life. These people are choosing to live closer to the action, their work, the restaurants they enjoy, and to be where their friends have moved. But they still want the homey feeling and not a sterile, boxy, closed-off place. They want flexibility to use the space in ways that work for their lifestyle rather than traditional layouts of a separate, clearly defined living room, dining room, kitchen, and bedrooms. Making an apartment feel more like a home can be challenging to you as a designer or builder. Homeowners between the ages of 45 and 64 tend to be at least 3 times more satisfied with their residences than renters in the same age group, Wylde's studies found.

Most boomer homeowners with families are used to three or more bedrooms. Apartments are a different matter. Over 90 percent of middle-income American renters live in one- or two-bedroom apartments. Wylde's study indicated that the number of bedrooms preferred was two or three rather than one or two, and the number of bathrooms preferred was two and a half, which is pretty unusual in apartments and condos. But currently, at least, they are not willing to pay for the extra bedroom and bath. New apartment dwellers really have to pare down when moving from a house. Generally, apartments don't have the extra nooks, crannies, closets, garages, and backyards that home do where homeowners they can stash their stuff until they need it (if ever), so that can be a shock for renters who previously had been living in homes. Typically, apartments or condos with extra storage space are hard to find and may charge a significant chunk of change for it, though the authors think that will change as stuff-heavy boomers pare down their possessions as well as demand someplace to put the things they can't live without.

Moving from a house to a multifamily home requires a change in mentality. Many empty-nest boomers *choose* to move to an apartment, which makes the transition easier. For those who do not want to live in an apartment but are forced to by economic factors or other reasons, it can be a little tough. As with everything, there are two ways to look at the differences, with a glass-half-full attitude or a glass-half-empty one. Apartment dwellers hear noises from their neighbors, whether it's music, a TV, rushing water, a vacuum cleaner, a barking dog, heavy footfalls, voices, or unidentified things that clunk, clatter, bang, or bump. It can either be an annoyance to hear other people's sounds, feel an invasion of one's privacy, and have to tone down one's own noise levels, or it can be a comfort to know that neighbors are close by. Giving up the yard can be frustrating to someone who likes to stroll or garden, or it can be liberating to the person who's delighted to give up maintenance chores. Parking a short walk away instead of right outside the door can be a daily irritant or it can make for an easy way to get a little more exercise.

This is where many people need a little help. Ciji Ware's book, *Rightsizing Your Life,* takes readers through the entire process of fitting their former life into less space. Other businesses have sprung up around the nation to help people decide what to keep, what to store, and what to throw away. One of author Kephart's neighbors has a business called "Bright Move," which helps with more than just packing and loading. Builders and developers can either develop their own businesses or contract with one or more in their area to make moving smaller a pleasant experience rather than a dread.

AGE-RESTRICTED, AGE-TARGETED, OR LIFESTYLE-TARGETED?

It's not that there's one kind of structure and layout for boomers and another for renters of all ages. It's that certain combinations of structures, layouts, amenities, and other features appeal more to boomers. Those same characteristics may also appeal to people in other age groups. As the authors have emphasized throughout this book, it's the lifestyle of the renters that determines what features they want, not necessarily their ages. That explains why several apartment and condo projects that the developer envisioned appealing to singles and young, childless couples just starting out also appeal to empty nesters, widows, widowers, and other older singles. If you want to target people with that lifestyle, one factor to include with the units is to "make them fun!" urged architect Ed Hord. Locating a project in the midst of shops, restaurants, movie theaters, dry cleaning, and other services, and within walking distance to transportation contributes to the fun factor.

Lofts or loft-like layouts are also fun and are sure to attract both empty nesters and young folks. Artisan Homes in Phoenix developed a loft project designed—they thought—for young urban dwellers looking to live in the middle of the action. They soon discovered that as many as half of their buyers were singles or couples over 50 years old. With a few minor adjustments in their marketing message, that number increased. It was the lifestyle that appealed to both the young singles as well as a significant segment of 50+ buyers.

Age-restricted projects don't have to be segregated. More and more intergenerational communities include an area or a building that is age-qualified. K. Hovnanian Homes built a condo section in a 5,000-home multigenerational planned development in Virginia. "Grandparents can be near their grandkids, go to their games, etc., but still be in their own enclave," said Mark D. Stemen, senior vice president of K. Hovnanian's active adult division.

CENTRAL CITY DOWNTOWN LOCATIONS

A significant number of 50+ homebuyers are moving back downtown. They are choosing contemporary lofts, more conventional condos, and apartments, but the common draw is the action, culture, sports, entertainment, restaurants, and the ability to walk everywhere they need to go in a single day.

Between 1990 and 2000, the population of 64- to 75-year-olds in downtown Chicago rose 17 percent. Austin, Texas, New Orleans, and Los Angeles have seen double-digit increases as well. A Del Webb survey of baby boomers indicated that 30 percent of those considering purchasing a home in an age-qualified active adult community prefer a community in an urban location. Urban living will not replace the Sunbelt as an attraction but there is plenty of room for both lifestyles.

Many boomer studies and surveys don't quantify what percentage of the 50+ boomers may prefer an urban center as a place to retire, but the National Association of Realtors reported that 12 percent prefer a city lifestyle. That report went on to say that family incomes play a strong role in determining where people choose to live. Boomers with household incomes over $100,000 are more likely to choose an urban area, the cost of living differences being a likely factor.

Nationally, the urban to suburban flight of all age groups may be too great for this 12 percent of baby boomers to offset, but on a local basis it's clear that the movement back to cities is a strong trend.

It may come as a surprise but Denver is the "baby boomer capital of the United States." Boomers comprise 32.8 percent of the population, according to the Downtown Denver Partnership, so an estimate that 50 percent of the downtown population is made up of boomers may actually be conservative.

Between 1990 and 2006, nearly 14,000 living units were built inside the Denver urban core. By some estimates, over 50 percent of those buyers are 50+ singles or couples.

That figure will remain an estimate, because builders and developers didn't track the age groups except to compute average ages in downtown. The number of families with children living downtown, however, was tracked.

The 2000 U.S. census reported a total of 13 percent of families in Denver having one or more children, and fewer than 5 percent of families living at the central core of the business district having any children.

Denver's growth in downtown living is indicative of a trend, or at least of a trend reversal. A study of 24 large American cities by the Brookings Institution and Fannie Mae found that a reversal of the trend of people moving out of downtowns was actually under way. The study concluded that the downtown population of Houston would quadruple by 2010, triple in Cleveland to 21,000 and nearly triple in Denver. Chicago was expected to increase by one-third to 152,000.

Be careful of taking statistics at face value. The Denver report cited a downtown population of 9,000, which is technically correct, but the area that most of us think of as downtown includes: "LoDo," the Central Platte Valley, Ballpark, Five Points, the Golden Triangle, and other center city neighborhoods. That area contains 83,000 residents, per the 2000 U.S. Census. No doubt the other cities mentioned have similar clarifications to make. Again, the local culture and accepted vernacular can trump what you read in a statistical report. It's important to understand these details before considering building in a downtown area in any city. And, one more point to put this mini-trend in perspective, the growth rate in downtown Denver is minuscule as compared to the suburban growth during the same period. Since 1980, the city and county of Denver's population grew by 12,000 people to a total of 500,000. During that same time, the suburban areas grew by one-third, adding more than 500,000 people. If such predictions of growth hold for other cities, a viable niche exists for the builders willing to go downtown.

That last sentence reminded Kephart of a story he heard about one of Denver's highly respected homebuilders who had just completed his first housing development in the city. The development consisted of 50 to 100 small single-family homes on an infill site in an old neighborhood—not much of a stretch, you might say. At a homebuilder's function, another builder asked, "John, why in the world did you choose to build in Denver and go through all that brain damage with zoning and codes?" John stared at the man and replied flatly: "Because that's where the market is."

When author Kephart realized that no building or community constructed in downtown Denver had been designed specifically for 50+ buyers, that spelled opportunity in his mind. Shortly after he formed that thought, he read an announcement for a planned senior living development in the downtown area. It's called the Cosmopolitan Club and it has a distinctly luxury hotel character which the owner, Balfour Senior Living, has set out to create. Robert A. M. Stern, one of the country's architectural stars, is designing the development. The project will contain 214 independent living and 50 assisted-living apartments. It's across the street from one of Kephart's favorite restaurants in LoDo (Zengo) and at the foot of Millennium Pedestrian Bridge leading to Union Station, the soon-to-be transportation hub for the metropolitan region. The Cosmopolitan Club will be a fine addition to downtown, but the door is still wide open for an active adult community downtown. Look at your own downtown for comparison; the numbers will vary, but the opportunity may be tremendous.

One caution offered by John D. Rhoad Jr., managing principal of the RMJ Development Group LLC in McLean, Virginia, developers of active adult rental communities, mixed-use projects, and intergenerational multifamily communities: "I hope the industry will recognize that if everything they're doing is age-restricted, the numbers alone don't earn the demand. It is a mistake to oversupply any particular product. As a rule, it seems like an idea comes and everybody's after it, but you don't want that to get overbuilt for a particular perceived demand and change everything you're doing. There's real demand but sometimes we overshoot it."

CONSIDERATIONS FOR CONDO/APARTMENT DESIGN

Let's have a closer look at the characteristics that boomers might find most alluring in multifamily living.

Respondents to Wylde's study indicated that renters wanted the biggest basic residence available. Size matters to them. But not much further down the list was a preference for a smaller, higher-quality home. That says that there's going to be a market for both! Respondents also said they wanted an enclosed garage, an accessible home, a great view, and a social program, preferably with fellow residents with similar interests. Wylde said the study showed all indications of a good rental market, "particularly in more urban, infill areas where there is higher density, more rental households, good amenities, and opportunities in the area." Condo owners are looking for similar advantages in their homes.

"First they're looking for location," said Evelyn Howard, president of research firm Howard & Associates, Bethesda, Maryland. "Many move to be near their children or their friends. But they're also looking for an environment where it's walkable, where there are lots of amenities, and a connection to the larger neighborhood around them. A major trend is to have a seamless connection to your neighborhood, living as a part of it as opposed to being isolated and separated. Those 55+ are not the pioneers. They like to follow when there is some vibrancy and life to it. After location, people pick out a particular building for the ambiance, the feel of it. Third are floor plans. Each of those elements is important."

Location! Location! Location!

A note of caution: A big mistake that developers, designers, and builders make is to decide to build an active adult project only because they can get it approved. Some local governments are fearful of overloading schools or tangling traffic if multigenerational projects are built so they only approve active adult complexes. If that's the firm's first entry into the active adult market, it could be a disaster for the project and all involved. Virginia condo and apartment developer John D. Rhoad Jr. warned, "The market needs to be the driver. It's wrong to go the other way. Then they wonder why it isn't working."

Just like with single-family home projects, those boomers who are looking to live in multifamily complexes are not interested in being put out to pasture, so a pastoral setting probably won't be nearly as popular as a project that's close to the action.

Urban infill locations or city centers are sure to soar in popularity as boomers enter the next phase of their lives. That includes condos or apartments above retail.

Though there are few tried and tested 50+ urban housing models to learn from, the fact is that older buyers and renters have always been a significant percentage of residents in high-rises and other building types found in urban areas. Elevator buildings offer some basic features that are highly appreciated by 50+ buyers, starting with a secure parking garage below the lobby level. The sequence continues as follows: a central secure ground floor lobby, often with a security person or doorman; the bank of elevators that serves the garage level, lobby, and all residential floors; and, finally, your front door. Security, convenience, and peace of mind are the qualities this experience provides residents. A common fitness center, indoor pool, multipurpose rooms, and clubroom can all be added at lobby level or in the penthouse to enrich the lifestyle.

Apartments and condominiums can be marketed the same way that Esprit Homes in Denver marketed their single-family homes at One Cherry Lane. "One-story living" was their headliner on signs, in ads, and in the sales office. The fact that there are no stairs to negotiate is not mentioned but is included in the lifestyle as described by the phrase one-story living.

Fifty-plus home shoppers now have another housing choice when looking for that new maintenance-free home. Many city centers now include a mix of housing options: lofts above stores accessed by elevators, freestanding townhouses, multistory apartments and condominiums, and maybe some single-family homes at the fringes of the center. The great thing about suburban city centers is the wide variety of stores, restaurants, food markets, movie theaters, fitness centers, yoga studios, and coffee shops that serve as community living rooms. There are also a multitude of services such as dry cleaning, shoe repair, and locksmith. You name it, a suburban city center may have it. They are much like a major downtown, but with parking.

Belmar, Continuum Partners' development in Lakewood, Colorado, is a model that is being closely watched by cities across the country. A well-designed and planned suburban city center like Belmar offers a complete lifestyle with a mix of housing and prices. Many 50+ homebuyers will choose Belmar because it's close to their old neighborhood. Many others will come who may have considered downtown, which is also rich in housing choices for boomers. Many smaller builders are involved at Belmar. Each builder is building for a particular market, price range and in a particular building type, to avoid too much head-to-head competition within the community. One builder is building three-level townhomes with attached garages at the first level, priced in the $400,000s. Another is building single-loaded, stacked flats in four-story buildings, priced from the $200,000s. Parking is in detached garages and open parking lots. Others are building lofts above stores, priced from the $500,000s. The developers are also planning some other apartments and home types.

Amenities

According to Wylde, the top 12 amenities desired by people 55+ who were moving to multifamily residences are, in this order:

1. Sidewalks
2. Mature trees
3. Green/open space
4. Walking trails
5. Upscale landscape
6. Indoor pool
7. Dog park
8. Outdoor walking trails
9. Outdoor pool
10. Upscale entrance
11. Clubhouse
12. Fitness and wellness center

A note of caution: Developers or designers with multifamily experience may think they can transfer their experience and skills into the active adult market. Not true, declared multifamily developer John Rhoad. "The truth is, it's really not the same product. Yes, it's stick and bricks but it's not a direct application of talents learned in the non-age-restricted projects. They fall down in pulling all of the amenities together. They don't know how to provide the right amount of socialization opportunity, what the amenities should be and how they should work. An age-restricted building with no amenities or the wrong ones will not work!" K. Hovnanian's Stemen added, "Pick an architect with active adult, multifamily experience. Remember, if you make one mistake, you haven't just made one mistake, you've made 166 mistakes" or as many units as you have in the project.

Condo owners only own the property within their wall space, with all homeowners jointly owning the rest of the property, including the amenities. Stemen urged hiring a quality condo manager early on to help organize and manage the all-important home owners association (HOA). "Then there won't be a bunch of surprises for homeowners later on. Assessments are not well received."

Clubhouses and Club Rooms

Even though Wylde found clubhouses to be the bottom preference out of the top 12 amenities, they can make or break a complex. Even if a bad one doesn't break it, a good one can make the complex a big success. Possibly the reason it was ranked so low was that people who responded to the study had poor previous experiences with clubhouses. Some clubhouses are virtually unused. That can be for any of several reasons. It may be inconvenient for residents to get to. It may have poor ambiance. It may not have a layout that works. It may not have rooms or facilities that work well with the activities the residents want.

One question for a developer or designer to answer is whether to have a separate building for the clubhouse or put it inside the building. "There is a divide in the industry as to what the role is in building the physical plant so it facilitates interaction and camaraderie," said Rhoad. "A common element (or building) outside is one way of thinking. I like integrating the social area within the building."

Author Schriener lives in one of RMJ's apartment complexes, The Evergreens in Columbia, Maryland, a 156-unit, 55+ age-qualified apartment building four stories high, where the center of the building on the first floor houses "the club room." That 3,200-square-foot space consists of a catering kitchen equipped with a refrigerator/freezer, microwave, sink, and lots of counter space; a large room in the middle with a fireplace and conversation area with sofas and chairs, several tables and chairs; a semiprivate area behind the fireplace with more tables and chairs; an area with a piano, a large-screen television, books in a library style and comfortable chairs; and a multi-purpose room with built-in shelves, a chair, a sofa, and a pool table. The space backs up to an outdoor multipurpose area with barbecue grills, a green area, tables, chairs, benches, and the pool and hot tub. Around the corner from the club room is a fitness room equipped with a handful of treadmills, elliptical machines, a recumbent bike, a universal gym, and weights. All of this is right across from the main entrance to the building, the leasing and management office, and a more formal sitting room.

The club room is the heart of the complex. Residents pass by it often—when they go to the office, check their mail, walk their dogs, go to the pool, buy something in the vending machine outside the fitness center, or even come into or go out of the building if they don't park underground. There's usually someone in there somewhere doing something, and often people will just go sit in the conversation pit and wait for someone else to come along. Pretty soon a small group has gathered and they are chatting away. It's also where the social events are held—the monthly birthday party, pizza and ice cream socials, pot lucks, concerts, monthly afternoon teas, daily continental breakfasts, weekly poker games and other games, the various special-interest club meetings, New Year's Eve and Halloween parties, and private and semiprivate events hosted by residents. After a pipe broke and flooded the club room, the floor buckled and had to be replaced, necessitating closure of the club room for weeks. Residents walked around morose, aimless, and complaining until it opened again.

Whether the club room is successful because of its design and location or because of the lucky mix of residents who are sociable, eager to help, and easy to get along with is anybody's guess. Schriener suspects it's both.

The main thing to keep in mind when thinking where to put the club facility is that it's a place to gather. "It's a place for spontaneous gatherings, a place where people can hang out," said researcher Howard.

Condominium- and homebuilder Jim Chapman, who heads Atlanta-based Jim Chapman Communities, said, "I build the clubhouse first, as quickly as I can get the permit." He holds meet-and-greet events there, and early on integrates current resident "ambassadors" with prospects in order to put names with faces. "The clubhouse has been a vital community hub for us," he said, and added, "I won't do a 10-acre, 40-unit active adult community because I've got to have a clubhouse but can't justify it for that size project. A clubhouse is just the price of admission."

In condo projects K. Hovnanian's Mark D. Stemen also insists that it's a tremendous advantage when selling the units if the clubhouse is built first, even though it's expensive. "Also get a lifestyle director or concierge on board early if you can," he said. He figures the rule of thumb for clubhouse sizes for condos to be approximately 25 square feet of clubhouse per home. "Up it if you have a small number of homes, lower it if you have a lot of homes," he said.

Part of what makes designing a clubhouse or club room challenging is that one size doesn't fit all. What may appeal to one age group or those with one lifestyle does not usually satisfy others of different ages or with different lifestyles. "There's even a significant difference between the [age] 50 to 60 demographic," said Rhoad. In multigenerational projects, it's hard enough to try to appeal to various interest levels, but even in active adult complexes that have residents' ages spanning 30 years, you have to cater to two or three different sets of preferences, some of which may be at odds with each other, in one facility. Some of the solution lies in making the clubhouse or club room space flexible enough to accommodate a wide variety of activities and some of it lies with the complex after the project is finished and how it manages the social aspects of the living experience there. However and wherever the clubhouse or club room exists, for it to work for the residents, said Rhoad, "it's all about interconnectivity."

Pools and Fitness Centers

How you treat the pool and the fitness center in a complex is different when you are designing for the 55+ population versus for the intergenerational crowd. "You want to have privacy!" declared Manny Gonzalez, principal in the KTGY Group Inc., a Santa Monica, California, design firm. "In other audiences, you want it up front to show it off. In rental facilities for 55+, they want privacy and convenience. They aren't looking for their date outside the gym like those living in general audience apartments might. And, residents shouldn't have to cross parking lots to get to the facilities."

Chapman suggested taking a very simple approach to fitness centers: a 15 by 20-foot room with televisions on the wall, hand weights, bar bells, and four or five pieces of good-quality cardio equipment.

Indoor swimming pools and hot tubs or therapy pools are expensive but very popular, said researcher Evelyn Howard. An alternative to providing a fitness center on site is to affiliate with one nearby, hopefully *very* close by, she added.

Business Centers

Apartment buildings built in the last decade or so were proud of their business centers, which were usually closed-in rooms stuck away somewhere in the building with a computer or two or more, a printer or two, and an Internet connection. They were designed for work, not for socializing. Those days are gone.

Nearly all boomers who are potential customers for the condominium or nonsubsidized apartment market have computers and Internet connections these days. If for no other reason, boomers want to be connected to their children or grandchildren, and that's who got them onto the Internet even if they'd been laggards before that. That doesn't mean there's no need for a room in the clubhouse or club room with computers, printers, and Internet connections. You just don't need a bunch of them like you would have a decade ago. Many good, solid printers are available now with extra capabilities such as faxing and scanning, which makes more services available to residents for very little extra cost to the complex. You could design that room as much for the social considerations as the technical ones. So you could place it near the action rather than segregated from the rest of the social environment.

The computer room in author Schriener's apartment building is equipped with two Internet-connected computers with large screens, two office-quality printers, one with faxing capabilities, a flexible workspace, built-in comfortable seating, and book-shelves stocked with a variety of coffee table books, novels, and classics. The room is separate from the club room but is right off the main hallway, across from the hall-way entrance to the club room kitchen, near the main entrance. Residents pass by and often stop in to talk to whoever is in there. The door—with a window—can be closed but nearly always is open. The location and set-up are conducive to a good balance of privacy and quiet for working, and opportunities for socializing.

Apartment/Condo Elements

No Wasted Space
The popularity of "great rooms" that encompass several uses, such as family, living, nook, and kitchen all in one well-defined space, is no different than it is in single-family homes. (See Chap. 6.)

Give boomers a choice between a lovely floor plan with space they can't use and a sim-pler plan that gives them every available square foot to use and the authors bet many will choose space they can use. Author Schriener had her choice of 2 two-bedroom layouts when she was looking in the building in which she currently lives. One was 160 square feet bigger and was on a corner, which would eliminate some neighbor noise. But she saw a lot of wasted space taken up by the foyer and hallways, plus the rooms were laid out such that she thought there was less usable space overall. So she chose the smaller apartment and has been quite pleased with her choice. The important things to note here are that she chose flexibility and usability over space, and it was her *perceptions* of what the space was like that tipped her over to choose the one she did.

Maximum Use of Available Space
Boomers want it all, so it will be a challenge to give it to them. They want a great kitchen, a minimum of two bedrooms and preferably two baths, a balcony or sunroom, large clos-ets, a good view, extra storage, and parking. That means squeezing out all of the wasted space you can. In apartments and condos you can skimp a bit on the size of the kitchen as long as the quality is high. Balconies take up space that can't really be used. "We found that the balconies for the senior buyer are underutilized, so we chose to bring that space inside the unit and turn it into a den or sunroom" in the age-qualified condos within the intergenerational community in Virginia, said K. Hovnanian's Stemen. Most multifamily units still have the outside balconies.

Flexibility
The concept of flexible use is more difficult in smaller living units if you're thinking at the scale of a room. Rooms can be used in any way imaginable if they are properly located relative to baths, the kitchen, and the primary living spaces. A bedroom may be

used as an office, dining room, nook, or entertainment center. Almost any room can become anything else, but in a two-bedroom condominium, chances are that the larger bedroom will remain just that, as the living room will not change use and it's likely that only one dining space exists. So, beyond the ability to use the second bedroom as a den or a dining room, where's the flexibility? This is where people can get creative, especially if the rooms offer the opportunity for shared use. A smaller, "odd" corner or niche can become a built-in office that can be closed from view with doors. A nook or dining room can double as a library with floor to ceiling shelving full of books, which is exactly what author Kephart did in his dining room. Those odd niches in rooms are great opportunities to use as hobby stations or display areas. The authors have seen them used as an aviary, a reading hide-away, an aquarium, a family photo and trophy space, and an ad hoc bedroom for guests. Kephart's wife's mother regularly used a small corner in her bedroom as a reading spot. Author Schriener once lived in a New York City apartment where the unit across from hers had a bathtub in the kitchen, which the tenant put a large board over and used as a desk. Look around your home and imagine how one of those corners, bay windows, low ceiling space under stairs, or gable dormer spaces could be used. Typically, living and great rooms double as dining space, entertainment rooms, and all sorts of family activities. Kephart's mom and dad literally moved the furniture out of their small living room, rolled up the rug, and danced with their friends on happy occasions.

Lots of Light

Light is a happy condition bestowed on the world or within our homes for only the cost of allowing it in. Sunlight is the best of all. It can actually nourish us with vitamin D and lift our spirits by just being present on our face. Without abundant light we can become depressed and moody, especially after a long winter of being locked indoors away from the cold and the sun.

Nearly all of the beneficial ingredients of sunlight can be duplicated with artificial light today, but it is seldom applied in the quantities needed for therapeutic effect.

Artificial lighting that allows for at least three levels of light is required: one for cleaning, other household work, and to boost the spirits; a bit less for normal every day use; and a third, lowest level for entertaining or relaxing.

Add a minimal amount of artificial lighting with bright spots at work places in the kitchen and baths and perhaps some designated reading areas. This level of light is inadequate for cleaning though. A much higher level of light is needed to see in those dusty corners and the fact is it takes more light when you're 55 than 25. Just try to read that restaurant menu by the romantic light of a flickering candle. It's all reduced to guesswork at that point.

Universal Design

Universal design (UD) emphasizes user-friendly design for everyone with the goal of accommodating the needs of people of all ages, sizes, and abilities, including the changes people experience during their lifetime. These features should be seamlessly integrated into every home's overall design.

The benefits are a more comfortable, safe, and user-friendly environment for everyone from children to seniors. UD is especially valuable as we age and aren't quite as strong as we used to be, aren't able to bend quite as easily, can't see as well as we used to, or our balance is a little shakier than before. Perhaps an accident has left us with limited mobility. UD provides ease of use and maximum comfort for everyone, not just those with disabilities or special needs. For example, everyone can benefit from a stepless entryway, whether you use a wheelchair, push a stroller, pull your wheeled luggage, or carry groceries. Adjustable counter heights in the kitchen or pull-out workspaces are not only great for children to use safely without needing to stand on chairs, but are much more comfortable for those all day cooking sessions during the holidays when you can sit and prepare food at the same time. Although we may be the most technologically advanced society in the world, we often find ourselves having to constantly adapt to our homes rather than having our homes adapt to us. Universal design seeks to reverse that experience so the more UD features you can incorporate into condominium or apartment spaces, the better. (See more about UD in Chap. 4.)

Green, Green, Green!

Condominiums and apartments leave less opportunity for green construction than do single-family homes. You can't orient every unit the best way to maximize natural daylight or warmth from the winter sun or protection from the summer sun the way you can a house. Competition often is such a factor that green features go in order to make the complex competitively affordable for prospective occupants. So the responsibility lies as much with the resident as with the builder.

The pressure to go green is on. You can't escape it today: The neighbor who lords it over you because his car is more fuel-efficient than yours. Author Kephart feels so guilty about his gas-guzzler car that he has begun sneaking looks at "smart cars" on the street, but he's got that neighbor in other ways. Kephart has photovoltaic solar collectors on his roof, ultra-efficient furnaces, on-demand water heaters, and his house is correctly oriented for maximum use of the energy from the sun. Take that!

The point is that there are endless ways we can all be conscious of the greenness of our lives but we each may choose a different path to being green. We can recycle everything, grow our own vegetables, and walk to work or ride a bicycle when the weather permits. We can move to a smaller, greener home, but what about the people who buy our old energy-eating house?

California is leading the way again with more regulations on green home design than anyone else. Cities, counties, and whole states are all jumping on the idea of regulating us to make us do the right thing, according to them. We should all do our part but if we only walk or take the bus or light rail and never drive a car, couldn't we get credit for that when we build our house? Maybe we could get 20 points toward LEED (Leadership in Energy and Environmental Design) certification. Kephart's wife should get at least 5 points for recycling, and they deserve 10 or 15 points for planting trees that shade the south side of their house in the summer. Her car gets over 30 miles to a gallon and they tend to choose it over his so they should get another 5 or 10 points there. Whoops, they do fly to other cities a few times a year and once overseas somewhere, so they should be docked points for that.

It's nice to have an energy-efficient home (see Chap. 4), but our homes needn't carry the whole burden of energy efficiency by themselves just because they are so easy to regulate. The truth, in the Kepharts' case, is that they will never get the money back that they spent to save energy. They could afford it, a prerecession thought, but not everyone can afford homes that cost more to save energy and perhaps see their money back in 7 to 10 years. Maybe Kephart's neighbor should be docked 10 LEED points for how many times he flies to New York on business. He may have to add insulation in his walls to compensate for it. That will show him!

Storage

Boomers have lots of stuff. They are reluctant to part with it. So they need somewhere to put it. "Storage is the very key to this population," said researcher Evelyn Howard. Storage works best if it's not a last-minute add-on; factor it into the design of the building itself, she suggests. Don't fear that the storage space you add will not be appreciated. "There is never going to be *enough* storage in any of these units," said architect Manny Gonzalez.

Some ways to incorporate storage within the units themselves are

- Add a few square feet to closets and laundry rooms.
- Add closets in hallways, bedrooms, and bathrooms.
- Add a separate storage room.
- Add extra cabinets in the kitchen and laundry room.
- Allocate some of the patio/balcony space for storage.

Some ways to add storage elsewhere are

- Create a storage room with large lockers or cages for each apartment/condo. This could be on the ground floor, in a basement, or in a separate area in an underground parking area.
- Put storage cages at the end of some or each parking space in underground parking.
- Extend the length of enclosed garages.

Parking

The 55+ persons who are moving to a multifamily unit want enclosed parking spaces, Wylde found in her study of middle-income renters. Ninety percent or more deem them important, very important, or essential. About half of the people who desired to live in spaces over 1,500 square feet wanted two spaces, versus 38 percent of movers who wanted to live in spaces under 1,500 square feet.

For author Schriener, enclosed parking was one of the deal-makers for her choosing that complex. Ample open outdoor parking was available for free, but she chose to pay for secure underground parking. It enabled her to feel safer on those occasions when she returned home from a class or outing late at night; it made carrying groceries and other packages easier because it's closer, more temperature-controlled, and drier; and, very importantly, it saved her from having to dig her car out after snow storms or get into a boiling hot car in the heat of the day in summer. Very few other apartment complexes in the area offered that parking amenity.

"The Artisan Lofts on Osborn" in Phoenix are clearly designed for a mixed market. Elevators serve all units, but some have two stories of living space. The single-level homes appeal more to 50+ buyers, while the younger buyers are perfect for the multiple-level homes. Any buyer, 50+ or not, can choose either, of course. These types of buildings can offer specific designs and features for each buyer segment and create a better product for everyone. The plain vanilla floor plans that result when trying to design for everyone are not appealing to anyone. The Artisan Lofts were designed for a traditional loft market and were surprised by the number of 50+ buyers that materialized. "No walls between you and your imagination" was the marketing slogan. Offering wide-open spaces a couple could finish as they desire had great appeal to the hands-on consumer who wanted to be involved. This strategy was a great success at the Artisan Lofts and is a strategy to consider when designing any home for the 50+ buyer.

Wrapping It Up: One Builder's Secrets to Good Condominium Design

Chapman said he got to be bigger than any of his competitors other than Del Webb when measured by number of closings by sticking to the following principles when building condominiums:

- Stay with one-level living.
- Make it an easy-living home. Make the doorways wide, have no short toilets, and adhere to universal design principles.
- Design with roof trusses such that almost every room has a vaulted ceiling.
- Let in plenty of light. Install wide, tall windows. "Whether you're 55 or 80, you don't want dark spaces," he said. It also gives the illusion of a larger space. "It's not meant to be a trick; it's just meant to be better."
- Give attention to outdoor spaces. "None of our competitors do outdoor living worth a hoot," he said. His entries have 8-foot-deep entry porches and French door access that can be turned into a screened porch. A lot of people who downsize are giving up outdoor spaces and appreciate the effect.
- Give each unit a two-car garage.
- Keep units maintenance-free for owners so they can lock and leave if they want and enjoy the benefits when they are there.
- Provide adequate storage for each unit, something condominium buyers often have to give up. He adds a climate-controlled bonus room above the garage with a permanent staircase in the back of the kitchen. The room can be unfinished for storage or turned into a finished room for an office or guests. "Before I came up with a bonus room, people either rented a storage unit or didn't buy from me. I'm sure it was a deal-killer for some," he said.

Chapman's business model is based on doing infill communities, 20- to 50-acre sites in a very tightly populated section of metropolitan Atlanta outside the main beltway. "You want to be in those densely populated areas because you want to draw enough people in a 10- to 15-mile area who want to downsize and who want to stay in the area rather than go another many miles away to Del Webb," he said.

URBAN HOUSING MODELS

As we pointed out earlier, older buyers and renters have always been an important component of the market for high-rise elevator buildings. Security, convenience, accessibility, and amenities are all important reasons.

Individual storage rooms at the garage level will answer one common objection to condominium living—inadequate storage. From this basic menu of features and services, the design vision can look further for a concept that transcends the physical elements to create a design that appeals to people on an emotional level. The five-star hotel concept can do that for an upscale market. Five-star hotels have a reputation of service on an exceptional level. There are a few of these "hotels" in Beverly Hills. An essential component of the idea is the inclusion of hotel services into a residential community. The current examples operate as hotels and sell a percentage of the suites, thus assuring that the service level will not falter.

An affordable vision could incorporate a "lifecenter, fitness and health concept," as the defining feature (see Chap. 3). This center, open to the general public for memberships, could serve the entire downtown market. Homes could be built above and around the lifecenter since the fitness areas require few windows and can be constructed on several levels.

"Mixed use" is a foreign concept to most suburban builders, but may be required in many urban developments (see Fig. 7-5). Both the "luxury hotel" and the "lifecenter" concepts have mixed use as an essential ingredient. Partnering with someone with commercial experience is one way to gain that expertise quickly with the least commitment if you choose to "test" the downtown market.

This book describes several 50+-community types currently found around the country. Nearly every one of them, at their essence, has the potential building blocks with which to build a vision for a downtown building or 50+ communities. A downtown housing development can even take its inspiration from a small town, as Eco Village has done in Los Angeles. They can certainly be either: exclusive; a co-housing community; alternative; or spiritual. Downtown homes can also be "reverse second homes" for those who have moved to the Sunbelt but want a place back home to use when visiting family and friends.

The flexibility of building with wood-frame construction is severely limited, though not prohibited, in downtown zones. Much can be done within the four- to five-story limit, but many opportunities will require building higher which will change the way you build. Associating with an experienced urban partner may be a good way to start your downtown business. Some general contractors may make a good partner. They know how to build in the city and you know what buyers want in their home designs.

Figure 7-5
The Crest Apartments in Denver located its clubhouse above a retail center facing the street. (*Photo courtesy of KEPHART.*)

SUBURBAN CITY CENTERS

Suburban city centers are a new/old phenomenon similar in origin to traditional neighborhood developments. Many suburban communities that have grown as bedroom satellites to the central metropolitan city want to establish their own identities. They have looked around their city and realized that there is no place that personifies their community, no place that people gather for special events, no place that residents show off with pride to visiting friends and relatives.

Planners looked to the past for models of city centers and found nothing to surpass the traditional main street or downtown of the typical American small town. Downtown was a place of business, government, and an activity center for residents. A park or square was often the focal point with a gazebo for bands to play in and shade trees to sit beneath. The downtown square was the place to be; it defined the community. Houses were right around the corner and often apartments or offices were located above the stores and restaurants.

Belmar

Belmar, a redevelopment of an old suburban mall west of Denver, now provides a safe urban lifestyle in the suburbs. While there are no age-restricted neighborhoods within this development, a significant proportion of homebuyers are over 50. All housing in the center is made up of townhouses, lofts, condominiums, or other multifamily configurations, many of which are made more accessible by the inclusion of elevators. These compact building forms use less land than houses and are more compatible in scale with the commercial buildings.

Near downtown Denver, developers in the City of Lakewood spent a couple of decades bringing their center together. In Lakewood's case, two factors worked strongly in their favor. They had bought a large piece of land uninterrupted by streets or roads, and they had assembled all of their government functions there in various contemporary buildings. The lowlands were developed into a wetlands restoration area and Heritage Center. The properties at the edges of the two bordering arterial streets were reserved for commercial development and multifamily housing. Today the public buildings, surrounded by public plazas and open spaces, are the focus of the center, but the best was yet to come.

Continuum Partners bought a tired suburban mall, Villa Italia, across the street from the Lakewood Town Center, and worked with the city to create a fresh new development concept that finely mixed a variety of uses into an extension of the town center. Today, there are two major food stores, Whole Foods and Safeway, both in full operation. There are offices above retail, lofts above stores, townhouses, condominiums, apartments, a wide variety of name-brand retail outlets, and restaurants galore, which are all packed every evening. A Cineplex is the centerpiece in the heart of the center, across the street from an ice skating rink. Belmar and the Lakewood City Center is the place to be in Lakewood, something no residential community alone can create.

Author Schriener lived in a wonderful apartment complex overlooking a delightful little pond right next door to Villa Italia in the late 1970s. Even by then, Villa Italia was just another shopping center. It was in no way new or innovative and never really fit in with the newer, more exciting neighborhoods that were springing up as suburban infill communities nearby. Schriener and many of her neighbors largely ignored it, opting instead to shop and dine at trendier, more exciting malls several miles away or the staid but reliable stores and restaurants in downtown Denver. When she went back to visit over a decade later, she was shocked at how shabby the shopping complex looked and how outdated it seemed. Her mom, who lived close by on the other side of the mall, refused to go there, citing it as unsafe. It was a big, ugly dinosaur, and it certainly wasn't an economic boon to the area. But look at it now!

A well-designed and planned suburban city center like Belmar offers a complete lifestyle with a mix of housing and prices. Unlike a major downtown, the housing types found in a city center can include wood frame construction types. Housing above stores or offices may require fire-resistant construction, but freestanding condominium buildings and townhouses, up to and including four stories plus parking below, can be wood-framed under current codes.

Cities across the country are attempting to create or build upon their current centers to include government, offices, retail, entertainment, and housing, all within an environment of activity that will draw people there to enjoy themselves and maybe do a little shopping. Santa Barbara, California, architect Barry Berkus calls these places "fun zones." Housing within these zones and adjacent to them will take on new forms as the idea grows in popularity. It will be higher in quality and design and will include housing types for everyone, rich or poor, mixed together.

Suburban city centers are springing up for a variety of reasons: the old cities have spread out to as much as 3 times their post–World War II size while simultaneously losing 30 percent of their population. Cleveland, Ohio, is a classic example of this phenomenon. The increased distances of travel to the urban center from the edges has revealed the need for urban services and entertainment opportunities closer to homes. Suburban communities have also awakened to the realization that their city has no identity, nothing to capture the imagination.

Town Centers

Columbia

A 14,000-acre planned community called Columbia arose between Baltimore, Maryland, and Washington, D.C., in 1967. It was heralded at the time as bold and innovative for its planned diversity, village-based livability, and integrated open spaces. The Rouse Company was the developer, headed by Chairman James W. Rouse. Columbia today is composed of nine villages and a Town Center, boasts nearly 100,000 people, more than 5,500 businesses, 27 public schools, nearly 4.8 million square feet of retail space and 5,300 acres of open space, including parks, lakes, trails, and playgrounds. More than 60,000 jobs support the local population. Columbia is a diverse yet upscale community. Average annual income is nearly $97,000. It has a wide variety of housing for homebuyers and renters of all ages but limited choices for the 55+ market.

> Author Schriener lives in Columbia, in an age-qualified apartment complex right next to the big mall in Town Center. It's the only age-restricted rental property in the vicinity. Columbia is its own nearly self-contained town. It offers a wonderful mix of suburban-style conveniences; lush, beautiful trees; and easy access to both Baltimore and D.C. But it's not walkable. Walkways were incorporated into the original design within each village, but residents are crime-wary and need more than the offerings of their own villages. Other than the 200-plus mall stores and restaurants, Town Center residents have to drive everywhere to get what they want and need, including basics such as groceries. A proposed modernization plan will go a long way toward revitalizing the Town Center into a suburban city center, a thriving community, and a very desirable place to live.

Columbia, at just over 40 now, is beginning to show signs of a serious midlife crisis. Town officials and planners recognize the need for Columbia to be modernized, and the town is looking at plans to remake the downtown area, known as Town Center, over the next 30 years. General Growth Properties (GGP), which acquired the Rouse Company, owns much of the land and until recently was leading the planning effort. GGP filed for bankruptcy in April 2009 after wrestling with $27 billion in mortgage loans, creating the biggest real estate failure in history and throwing development plans and schedules for Columbia into jeopardy. The Howard County Council voted in September 2009 to allow property owners in Columbia to submit redevelopment plans directly to the county instead of having to go through GGP (and Rouse before that) first, as has been required since the town's inception.

Exactly what that modernization will ultimately look like is an ongoing discussion, but it likely will include higher-density residential areas than currently exist, including affordable housing ("affordable" for two groups: people making about $80,000 to

$120,000 a year, and people making $80,000 a year or less); a handful of distinct neighborhoods; mixed-use areas mixed with open and green spaces; infrastructure expansion; lots of services and amenities; and abundant walkways, trails, and access bridges. The hope is that plans put on hold during the recession will proceed smoothly and with lots of community input.

It remains to be seen how much of the proposed 5,500 new housing units will be age-restricted or age-targeted. Even if that number is low, the area is bound to attract boomers because it has the elements of urban developments that attract them: proximity to retail, services, business, culture, and other amenities; availability of mass transit (lacking now) and proximity to highways; being in the middle of the action; and a revitalized, feels-like-new community.

Other Town Centers

The downtowns of small cities, sometimes with a square with a courthouse right in the center, like Kephart's hometown, Perryville, Missouri, were places where people gathered on special occasions. In Perryville, the parade for homecoming started in the square. The majority of businesses were located there and Saturday mornings in the square were always special. The farmers from around the county all came to town to shop and to sell their eggs or fresh milk and cream at the local farmer's exchange store. Kephart's dad ran that store and young Kephart did odd jobs like sweeping floors and bagging animal foods for the farmers, who brought corn, barley, wheat, or rye for the store to grind and mix with vitamins and molasses for their cows, pigs, and chickens. Everyone would meet at the Park-Et Restaurant for lunch, looking forward to a free afternoon. Downtown was the place to be, and that strong image, which existed in the minds of every citizen and visitor, is what suburban cities lost or never had (see Fig. 7-6).

Aurora, Colorado; Union Township, Ohio; Redmond Town Center, Redmond, Washington; and numerous other municipalities are working to develop their own town centers. In Aurora's case, the center has been in the works for 20 years or more. Today they have many of the ingredients assembled, but the mix of disparate uses all separated by busy roads makes any cohesive image difficult to create. The best prediction for the future popularity of suburban town centers among the 50+ is that they will be far more attractive to larger numbers of homebuyers than downtowns. Even more importantly, in their maturity after the city has done its part, a suburban city center location will serve a major niche that competing developments cannot match. Builders proficient in building any attached housing type can find a place within a suburban city center.

There is no sure bet in housing development, of course. Timing is always a judgment call. It takes 10 or more years for a suburban city center to take form and start to be recognized by the public as an important place. Belmar itself pushed Lakewood's center to the forefront. Even though the city had completed its part of the center, it didn't take on a regional image until Belmar opened for business. It's much like buying a development parcel from a master-planned residential community. The public image that a community creates is all-important to a builder, as is the time to buy a parcel while still affordable but

Figure 7-6
Downtown Madison, Georgia. (*Photo by Mike Kephart.*)

close to its peak of maturity. Suburban city centers still carry some risk for a builder, but compared to attempting to create a public image for your community on your own, the odds are substantially better.

REFERENCES

1. Birch, Eugenie L., "Who Lives Downtown," The Brookings Institution, Washington, D.C., November, 2005.
2. Del Webb, Survey of Baby Boomers, Pulte Homes Inc., Bloomfield Hills, Mich., 2005.
3. "Downtown Denver Build-Out Scenario Report," Downtown Denver Partnership, Denver, Colo., November, 2005.
4. Education Sessions at International Builder's Show, National Association of Home Builders, Orlando, Fla., 2008 and Las Vegas, Nev., 2009.
5. National Association of Realtors data, 2008.
6. Ryan, Jenna, "A Study on Internet Usage by Older Americans," AARP Washington, D.C., July 9, 2008.
7. U.S. Census Bureau data, 2006–2009.
8. Sheahan, Matthew, "Bankrupt General Growth to Pay Dividend," *Leveraged Finance News*, December 21, 2009, http://www.leveragedfinancenews.com/news/bankrupty_general_growth_to_pay_dividend-201053-1.html. Accessed December 26, 2009.
9. Ware, Ciji, *Rightsizing Your Life: Simplifying Your Surroundings While Keeping What Matters Most,* Springboard Press/Hachette Book Group USA, New York, January 2007.

10. Wylde, Margaret, *Boomers on the Horizon: Housing Preferences of the 55+ Market*, BuilderBooks (National Association of Home Builders), Washington, D.C., 2002.

11. Wylde, Margaret A., "Active Adult Housing Prospects," National Association of Home Builders Webinar, September 2009.

12. Wylde, Margaret A., *Right House, Right Place, Right Time: Community and Lifestyle Preferences of the 45+ Housing Market*, BuilderBooks (National Association of Home Builders), Washington, D.C., 2008.

13. Wylde, Margaret A., "The 55+ Middle American Rental Market" Presentation at Building for Boomers and Beyond Conference, National Association of Home Builders, Philadelphia, PA, April 2009.

8

The Design Process Step by Step

"I'd like to design something like a city or a museum."

—Actor Brad Pitt

DESIGN FUNDAMENTALS

This chapter outlines the design process. It's a bit of a return to fundamentals but it is important to do the steps in order and resist the temptation to jump ahead before you have all the right information. Let's look at the steps in the order they should be done:

1. Market research
2. Vision and community design
3. Community amenities
4. Architecture for homes

Market research is covered below to a degree, also in Chap. 3. The vision and community design also are covered extensively in that chapter. This chapter will mention those steps as part of the design process but focus in detail on the last two steps, community amenities and architecture for homes.

Market Research

In-depth knowledge of your local market is the essential first step in the design process. There are a number of ways to acquire the necessary information to be able to focus on a particular segment of the 50+ market. Professional market research companies can provide the most comprehensive package of facts and figures. Most importantly, they can define your market area or where your potential buyers currently live. They can define your buyers, their needs and desires, the price they can afford, and the number of people out there who are ready to buy a new home. Strategies for information generation may include home shopper interviews, focus groups, analysis of the competition, studies on job growth, and other methods. Market researchers can predict the

Figure 8-1
Walking trails are extremely popular and they can be many things for many people, from a form of exercise to communing with nature. (*Photo courtesy of Joe Verdoorn.*)

WALKING VARIETY

79% OF THE 55+ HOME BUYERS LIST WALKING AS THEIR FAVORITE FORM OF EXERCISE.

market size now and in the future. Builders can conduct this research themselves if they have professionals in the company qualified for this work or, as often happens, simply rely on their past experience in the area. This latter method could be called into question by prospective lenders and investors.

Whatever way you conduct market research, you as an architect or builder must decide how to use the information. At best, research of past experiences of competitors can only give you a picture of what worked and what didn't in the past. What your future homebuyers will respond to could be the same as in the past or it could be dramatically different. Builders, planners, and architects need better direction to design for the 50+ buyers of tomorrow. Boomers, for example, have entirely different ideas about how they want to live than their parents, who populated the lifestyle communities such as Sun City (see Fig. 8-1).

Buyer Preference Studies
Fortunately, there is good buyer preference research available through firms like ProMatura Group LLC, based in Oxford, Mississippi. Founder and President Margaret Wylde is a respected authority on the mature market and has written several books stuffed full with market research from national and regional studies. The authors have cited her research liberally in this book. However, she is not the only game in town when it comes to research; many experienced, qualified firms exist that can help you, and the authors encourage you to meet them at industry trade meetings and conferences, or just do a little research on the Internet or get a referral from another homebuilder (maybe one that doesn't compete in your area) and contact one or more. Don't do this by yourself!

Wylde and other researchers make every effort to look into the future by asking homebuyers such questions as "Would you choose to buy a smaller home rather than a larger one if the smaller home had higher quality finishes and materials?" If 75 percent of respondents to this question said they would choose a smaller home in that case, it is a clear indication of people's desire for better quality in their new homes and a willingness to sacrifice size to get that quality. How that would play out in practice will vary.

Note that just asking a question in a survey doesn't always get you a clear answer. Sometimes you have to ask questions a different way or dig deeper. For example, in one survey Wylde asked, "Would you prefer an environmentally friendly home?" Buyers responded positively but held back when asked how much they would be willing to pay to ensure that the home is environmentally friendly. Later questions uncovered how much they would be willing to pay up front for energy conservation, part of being environmentally friendly.

> Surveying buyers versus relying on data from the past offers a truer perspective on the future and can keep you from making costly mistakes.

Local Research

Global information is great to know. You can identify major trends and make general business plans, but market research must be gathered locally prior to starting any design work. Bill Parks, former head of product design for Pulte Homes' Del Webb unit and consultant to the industry based in Arizona, tackled the issue of the apparent geographical diversity in the 50+ market. He said, "The 50+ market is much more fragmented and diverse than builders would like it to be."

Parks gets to the detailed local market potential by dissecting the 50+ lifestyle into its components to identify niches and opportunities. He starts with the family composition of 50+ buyers in the area:

A couple—most common family   Quantity_____

A couple with a disabled child   Quantity_____

A couple with an aging parent   Quantity_____

A single person   Quantity_____

Two unmarried people   Quantity_____

A gay or lesbian couple   Quantity_____

He then divides them further by whether they are retired or still working:

Fully retired   Quantity_____

Semi-retired—work at home   Quantity_____

Semi-retired—work at an office   Quantity_____

Fully employed   Quantity_____

Parks continues by separating them into avocations or activities they prefer:

Play golf or other games often   Quantity_____

Work full-time   Quantity_____

Volunteer a lot/a little   Quantity_____

Learning   Quantity_____

and other similar questions.

In the market area for a proposed development, he assigns each 50+ person to the appropriate subcategories for the purpose of identifying the size of each group. He then works with the homebuilder to glean design strategies from the results.

Look at the information in your own market. For instance, if most 50+ people surveyed in the market area are semiretired or fully employed, you may want to choose a site close to a business center. If the majority are retired and want to spend their time in pursuit of advanced learning, you may elect to approach a local higher education institution to develop a relationship and perhaps purchase property on their campus for the development. Any number of strategies can be considered using this information. The survey process can go further into the detail of desired community elements, home design, ability to pay, propensity to move, etc.

Market Studies

Astute developers, builders, and designers don't flip to the "Recommendations" page of a market study and skip reading the details and "boring" data. They find opportunities between the lines and in asking the question, "What's not here?" For example, a market study recommendation may state that a majority of people surveyed want a golf course in the new community you are planning. If that majority is anything but overpowering, you may want to look at the minority responses to see if you can gain insight into other desires of your future homebuyers. Golf courses can be an expensive amenity. Are the people who want a golf course willing to pay the cost and are those that don't play willing to pay for the golf just to get the open space?

Author Kephart was invited, as the architect, to observe several focus groups of potential buyers of a new 50+ community to be located in the middle of a flourishing tech center. The project size was only 25 acres and 100 homes were anticipated. Prices would range from $750,000 to $1,250,000, which was determined through the focus group process. The focus group attendees, some 30 people in total, were of one opinion about the possibility of having a clubhouse. They did not want one! Their higher priority for a community amenity was open space, no mean feat in this four-homes-per-acre density of large one-story homes. Nothing in the previous market research or in the survey of competitive developments would ever have led to this conclusion.

Professional market studies for specific projects are essential. It would serve you well to appreciate guidance from the marketing consultants and value their opinions as important information in the process of developing a detailed design program. The marketing professional, however, is only one member of the project design team. The team leader, usually the builder or developer, should work with his or her design team to evaluate the marketing consultants' opinions based on information they may possess, gained through experience. The team should then work with the marketing consultant to draft the final guidelines for design of the homes and community. Finally, ideas and unanticipated opportunities often surface during the process of visioning and conceptual design. These should be evaluated by the marketing professional, along with the team, and the guidelines should be modified to reflect these new directions.

Some people ignore this process of evaluation and change but it is essential in team-building. Many of these early process steps help everyone involved to understand the project goals and allow them to provide input. Once invested in the ideas, everyone can work in unison to achieve the goals.

Market Research in Action

Colorado Land and Home Company, a former custom homebuilding company, had purchased a small parcel of land that stretched along a feeder road leading to a private country club and other high-end single-family homes. The land sloped steeply down from the road, offering distant mountain views. Existing zoning would allow some creativity in planning, but neighboring infill properties had chosen linear townhouse schemes to address the hillside condition. After reviewing a few conceptual approaches to the community design, it became clear that the jurisdiction was not going to approve anything but single-family homes for fear of arousing neighborhood opposition from the townhouse developments. The builder had done no market research and had not hired a marketing consultant. Based on the success of the adjacent townhouses, he assumed that he could repeat their experience by offering low-priced attached homes in a quality neighborhood and sales would come easily. That was all the market research done to this point. Kephart was able to help the builder come to the decision to contract for a market study. He was relieved to see that his client would be gathering some real facts prior to designing homes for this development. The researcher gathered a great deal of useful demographic information regarding market size, prices to anticipate, and the types of homes that people surveyed said they wanted. He then conducted two focus groups with potential buyers he had identified to test his preliminary recommendations. They learned that the core conclusions were correct. These were all 50+ buyers looking for a new home in their old neighborhood that would better fit with their new way of life. Some were moving down from larger homes and some from the same size or smaller but they wanted features like first-floor master bedroom suites and maintenance-free living. Every single person insisted on quality construction and materials. To them this meant stone or brick exteriors, tile roofs, one-level living, high-quality interior finishes, and handcrafted details. Kephart was able to assist in the summary of the findings and to help develop the design program, incorporating his opinions with those of all team members. The project got built and all the homes quickly sold in the $600,000 to $800,000 range.

On a larger scale, Del Webb involved all of its design professionals, including the architect, when developing its own market study for Sun City Huntley, west of Chicago. The design team, including Kephart, toured older neighborhoods to see where the potential homebuyers lived. They visited community centers, healthcare facilities, and fitness centers in those neighborhoods. They all participated in focus group "fairs," where they sought buyer preferences for various features and priorities in the home designs. The fairs were informal gatherings where prospective buyers went from booth to booth to build their preference list, and those in charge of those booths recorded results. At Kephart's booth he asked for choices between a larger master closet and larger master bath and for their preference for closet access through the master bath or by separate door from the bedroom. They learned from the fairs that buyers wanted basements, no-step entries, one-level living, storage on the main level, mudrooms separate from laundry rooms, and large entry closets and vestibules.

Del Webb was unfamiliar with basements and with the mudroom and entry closet ideas since they aren't common in Arizona, Del Webb's base of operations. All of these things, coupled with first-floor storage and the master suite preferences, required massive revisions to some of their previously successful home designs from Arizona which they had hoped to repeat. Kephart was able to revise some designs, but others had to be redesigned from scratch. The builder willingly agreed to the changes to accommodate their future homebuyers.

Market studies go on to include community design, features, common buildings, and amenities desired by a particular market niche. Builders should look deep into the market to discover the distinct qualities in a community that smaller buyer segments strongly adhere to.

Resist the urge to over-generalize by reading national data and pay close attention to the buyers in the local market area.

Community Vision

Market research is the first step in design. Creating a vision to guide the entire design and marketing team is a critical second step. Establishing a vision begins with focusing on a clearly articulated segment of the huge 50+ buyer group. The temptation is to try to include everyone in the vision and ultimately satisfy no one.

There are many ways to approach settling on a vision. In essence, it involves narrowing down "the world" to your individual community, what it will look like, who you want it to attract, what those people will want to entice them to visit your neighborhood and hopefully become a resident.

Community Design

You can find much about community design in Chap. 3. More follows.

Cluster for Affordability

Few good examples exist of communities that group homes in intimate neighborhoods or clusters; 50+ buyer responses in the consumer preference studies reflect this fact. Yet the concept reduces land cost, increases density, lowers sales prices, and conserves open space. KEPHART, the firm author Kephart founded, has had a far more positive experience than most with this misunderstood development idea.

The clusters at the Arbors in Brunswick Hills, Ohio, consist entirely of single-family homes. Some of these designs were also used in more conventional lot arrangements. In this case, the clustering helped the developer/builder, Parkview, achieve the maximum number of homes allowed on the site and preserve large areas of natural lakes, wetlands, and streambeds. In addition, several 3- to 5-acre parks were located throughout the community as neighborhood gathering areas and focal points. As a result, all homes in the Arbors are priced in the low $200,000s.

Some municipality zoning refers to conventional but smaller lots as "clusters." The terminology is thereby confusing to buyers, which may explain some of the apparent indifference or resistance to "clustering." Buyers must understand the value of the location or price that is created by clustering homes into higher-density enclaves.

Oak Leaf Homes at Grant's Settlement at Rivers Bend, Ohio, markets their cluster homes as "villages," emphasizing the village character of each eight to ten-home cluster combined with one or more other clusters to form small neighborhoods.

Product design offers cues for shaping new homes and communities. Boomers, not kids, are buying the retro cars, the new "Bug" by Volkswagen and the PT Cruiser. Boomers and Gen-Xers alike buy furniture and household accessories at Pottery Barn. Maureen McDermott, sales and marketing director for Oak Leaf Homes, has connected with Pottery Barn, famous for simple, understated home design, for the interior design of their homes at The Villages at Rivers Bend, Ohio. The Pottery Barn design team came up with a design concept for each home, based on the baby boomer buyers that the architecture and community plan were designed for. "The minimalist interiors allow the architecture to really show," said McDermott.

The community of The Villages at Rivers Bend is modest in size, situated on the bluffs above the Little Miami River north of Cincinnati in South Lebanon, Ohio. A public river trail system exists as a major amenity. Canoe and boat rentals are close by, as are bike rentals and shuttles for river boat trips. Additional walking and biking trails are planned within The Villages at Rivers Bend to connect to the Miami River trails. There are a total of 140 acres in the community and much of the area will remain as natural open space in order to preserve the existing hardwood forest leading down to the river. Bird-watching blinds and towers are located along these trails for avid bird watchers.

Three groups of homes are offered in the community: Fredrick's Stand: 47 single-family attached homes priced from $169,900 to $222,900; Grant's Settlement: 30 detached single-family ranch homes in unique cluster neighborhoods priced from $288,500 to $327,900; and the Homestead: 52 lots for custom homes. Home sizes range from 1,261 to 2,147 square feet in the cluster and single-family attached designs.

Mixed Markets

Seventy-five percent of those surveyed by Del Webb who were likely to move chose a community with all age groups. Only 7 percent chose an age-qualified community.

Eighty-five percent of Americans over 50 said they have no preference regarding the age mix in their community or prefer it to contain all age groups (37 percent), and only

12 percent of the over 65 population would prefer to live in a community where most people are their own age.

Many builders are currently capitalizing on buyers' preferences for a mixed-age community and specifically the 31 percent that will consider attached housing. Small multifamily buildings combining three to four homes per building can provide homes for young professionals and other market segments in high-end locations. Stacking homes in two- or three-story buildings can reduce land use by increasing density and thus lowering costs to the consumer.

Three- to six-unit manor homes can provide a higher-density mix of housing types without the appearance of high-density housing. Manor homes look like large single-family homes and fit right into infill locations in older neighborhoods, giving buyers an option to stay in their neighborhood and still get a more affordable maintenance-free home.

Community plans can offer a more seamless mix of home types as well as different home types within single buildings. Two-story family homes mixed with single-story homes, manor homes, townhouses, and other attached building types offer 50+ buyers a variety of home types, sizes, and prices. It can also broaden a builder's market while giving buyers exactly what 75 to 85 percent of them say they want—a mixed-generation community.

There are lots of ways to mix things up. Hidden Lake in Frankfort, Illinois, has estate lots, cluster single-family homes, conventional single-family homes, courtyard homes, rear-loaded townhouses, and front-loaded townhouses. This product mix provides a range of home sizes, types, and prices to encourage a mix of buyers from families to singles to retirees.

Georgetown 2 in Batavia, Illinois, offers four-unit manor homes. An upper-floor flat is perfect for young professionals and the first-floor ranch is ideal for those who prefer a home with no steps. Two main floor master townhouses will suit a variety of buyers.

The community plan in Painesville, Ohio, mixes lots and homes block by block in what some have referred to as a "quilting pattern." Individual builders can offer small homes, large homes, townhouses, or condominiums. The community is organized into four neighborhoods of approximately 200 homes each.

Proper community planning can incorporate a mix of housing types with affordability for all by locating the higher-density homes, townhomes, and condos close to a town center or another high-ticket location such as a marina, lake, or a special vista looking out to sea or toward a beautiful mountain range. The one-story homes can then be more spread out with larger lots while remaining affordable. The higher-density attached and two-story homes also remain affordable by sharing the higher-cost locations among more families. The developer must keep these costs in balance within the community and the high-draw/high-ticket locations must be first rate for the concept to work.

Community Amenities

Successful "lifestyle" communities of the past owe their success to the experiences they created for their residents. Boomers expect more than their predecessors did in terms of accommodations and experiences. They are looking for more than plain-Jane accommodations and will respond positively if you provide the ones that enable them to play and socialize within their community. As the changing face of seniors gets younger and healthier, if you assume that the amenities that have been popular with seniors for years will also draw boomers, you may be disappointed. If you keep in mind that boomers intend to remain active and seek out activities that challenge their healthy bodies, you will be on the right track. Home shoppers who express a preference for an active adult community are more active than those who wish to move to an intergenerational community, studies show.

The physical expression of the experience, usually of a resort or life as a vacation, is most profoundly expressed in the community design and particularly the features or amenities in the community. An amenity is, by definition, something that is pleasant or conducive of pleasant thoughts. An amenity is also seen as a luxury or a valuable convenience that enhances comfort and enjoyment. The character of the experience is created through the unique assembly of amenities in the community and the homes. Let's look at them individually.

Golf Courses

The golf course–centered retirement community may be a dinosaur. Even in existing golf course developments, studies show that only 20 percent of residents are actually golfers. Of households of age 45+ residents, researcher Margaret Wylde's survey figures indicate that just 1 percent wanted their homes to be on a golf course or have a view of a golf course. This varies a bit by geographic region. More people in the West and South would choose a home in a golf course community or a resort-style development than those living in the Midwest and Northeast. In all cases, the percentage of those choosing the golf course or resort-style community was less than 7 percent. Home shoppers looking to buy in a golf course development want more community amenities than those who are looking to live in a traditional neighborhood or gated community. Homes in golf course developments are generally worth more than equivalent homes in traditional neighborhoods, which would make sense, given the extra benefits they enjoy.

Over half of survey respondents said they wanted a view of green spaces or parks, mountains, or the ocean or other water bodies. Whether they had their own view of the green space or not, 70 percent of respondents to Wylde's study cited open and green spaces as important. Therefore, Wylde concluded, developers and builders could save the expense of a golf course by providing the atmosphere that a golf course community provides—that is, lots of green space, good views, and a social gathering place.

Clubhouses

Much has been discussed and written about clubhouses. The debate goes on. Build it early, build it late; segment it a lot, keep the space open and flexible; put the fitness center inside, have a separate facility or partner with one nearby; put it near the entrance to the community to show it off, put it in a more convenient place for the residents; and so on.

Figure 8-2

A clubhouse can be full of amenities and activities. At Venetian Falls in Venice, Florida, Centex elected to oversize the clubhouse to compensate for the lack of usable open space in the community plan. (*Photo by Mike Kephart.*)

The one thing that nearly everyone does agree on, however, is that every community that appeals to boomers, whether it's age-restricted or intergenerational, has to have one.

People want their clubhouses! In every survey every researcher conducts, it comes back that clubhouses are popular (see Fig. 8-2). They tie the community together. They are not just there for activities. One strong characteristic of a clubhouse is that—ideally—it brings people together. Social events and celebrations happen there. Spontaneous conversations occur there. Plans are hatched there. Fun is had there.

As Atlanta builder Jim Chapman, head of Jim Chapman Communities, said, "It's the cost of admission." He builds his clubhouses first, both in his condominium projects and single-family home communities. He makes use of it from the get-go for community get-togethers and for event marketing to bring together prospects and those that have already bought so the latter can serve as "ambassadors" for the community.

Some builders wait to build their clubhouses until enough people have moved into the development to make the investment into the structure worthwhile. Their thinking is that with few people using it initially it wouldn't be economically sound and it wouldn't be much fun for the sparse crowd to go there.

The biggest mistake a homebuilder can make, in the opinion of many, is to not build a clubhouse or build one that misses the mark. That goes for communities consisting of homes, townhouses, condominiums, or apartments. If you want to attract boomers, give them a place to hang out together.

One condominium builder in New Jersey was experienced in creating communities for families but had never done one that was age-restricted or even age-targeted for the 55+ market. The only zoning approvals he could get were for age-restricted multifamily units, so he built a multistory building like he would for families, didn't have a clubhouse or even a club room, and offered no other amenities either. It was just a building. It failed. No surprise there. Many clubhouses lie as fallow as a baseball field in winter because they are badly located within the development or are badly designed and unable to facilitate socialization or other activities the residents deem important. Either way, it's a lose-lose proposition.

In ProMatura Group's 2008 study of age 45+ recent and prospective homebuyers, clubhouses were among the top preferences. Respondents were divided about in half on whether they preferred a basic clubhouse for meetings and social gatherings or a more lavish clubhouse with card and game rooms, a theater, an art studio, and so on.

If you give residents a reason to go to the clubhouse, it will be a better amenity. Social events draw people, especially if they include food. If your fitness center is there, people can use the clubhouse before, after, or instead of a workout. Some builders put extra storage in the clubhouse building, which gives residents another excuse to go see what people are up to.

Not every community should necessarily have a clubhouse. One example is One Cherry Lane, an upscale neighborhood of 86 homes in the Denver Tech Center. Focus group members—who later became buyers—elected to have a small cabana and a pool instead of a clubhouse, and this is a community of $1 million homes. The word "clubhouse" connotes a luxurious facility and that may not be what your buyers need. A simple pavilion where they can gather on occasion may be enough. This demonstrates the need to know what your local market wants. Are buyers in your market more interested in an exercise facility? Call it a gym or studio and keep it simple. Buyers 50+ know how expensive common buildings can be to maintain and they will keep a close eye on the homeowner association dues to pay for that maintenance. Depending on your chosen market niche, a clubhouse is better called a community center or something that doesn't imply golf if golf is not on your agenda. A community building may be necessary, however, to provide the programs and amenities your market niche is seeking, and to give residents a place to meet and socialize with fellow residents.

Figuring what size to build the clubhouse often is a guessing game. You want a place that's not too big or too small for your community. Some builders are enlarging existing clubhouses or designing larger ones and adding activities to justify the expense in order to make them more attractive to home shoppers.

There are rule-of-thumb formulas for community center sizes but it is best to size yours according to the amenities your market niche wants. The Arbors at Brunswick Hills in Ohio is a 200+-home, non-age-restricted community that has a 4,000 square foot clubhouse with an exercise room, billiard room, card room, and lounges. That's working well and the ratio of clubhouse size to the number of homes is 20 square feet per home. However, that community follows very closely the "lifestyle" community approach, or "nirvana," as Kephart has named it. An activities director is on staff conducting exercise programs, trips, competitions, and social gatherings. It's like a small Sun City without the golf. If your community vision is a TND (traditional neighborhood) where fewer than one in five residents say they want a clubhouse, you may elect to forego a community center altogether.

The community center can be segregated as follows:

Small community center: The typical minimal community building consists of a small meeting room or lounge that can be used for community gatherings or rented to residents for special occasions. The development costs may be better budgeted for walking trails, open space, and other inexpensive outdoor features buyers want.

Moderate community center: Small "lifestyle" communities can use this model. It adds an exercise room and outdoor pool to the minimal version above.

Large community center: The large "lifestyle," age-qualified community model depends largely on activities and amenities to attract buyers, so a large community center is necessary to house everything from golf shops to restaurants to fitness and wellness centers.

Centex Homes in Venice, Florida, chose to provide a larger-than-expected clubhouse in order to compensate for the fact that they had no golf courses in their Venetian Falls development. How many amenities and how large these clubhouses are is part of the strategy for creating an experience for the residents.

Near Orlando, Florida, Avatar's Solivita community has an urban street scene as its clubhouse-equivalent. Each function, such as fitness, the arts, a coffee shop, convenience store, and restaurant is in a separate building on this "commercial" street. It's a charming and well-utilized environment that goes beyond the concept of a stuffy clubhouse and turns it into a playful "experience." The only downside is that with the functions separated, fewer opportunities exist for residents to spontaneously interact. But the approach is popular with residents, and it makes the whole experience of living there more "normal" and less "retirement-community"-like.

Walking Trails

Topping the list of preferred amenities in ProMatura Group's study of the age 45+ homebuyer market was walking trails. That has held true since ProMatura's 2002 study as well and those of Del Webb. Don't look for that to diminish as boomers increasingly ease into the 55+ age group. Walking is so highly rated because of its low impact and the fact that it can be done with friends or a spouse.

You will do well to mix trails for walking and exercise with trails for strolling, socializing, and stopping to enjoy nature. And, what about when walkers share the trail with bikers, in-line skaters, or skateboarders? Yes, people over 45 engage in those activities. Incompatible uses diminish the quality of the experience for everyone. It only aggravates one group or another if you don't accommodate all.

A large city park in author Kephart's neighborhood segregates walkers from runners and cyclists. Skaters share the road with bikers but there's a limit. Runners follow a narrow dirt path around the park's entire perimeter in a one-way pattern to save space and avoid conflicts. Similarly, both bikers and walkers inside the park must travel one way only.

Most people walk for the exercise but some do it as a social experience. Kephart's wife walks with friends a few times a week, followed by a stop at the local coffee shop for more conversation and to meet other friends. A truly rich environment of trails, walks, and paths should be provided in a 50+ community to encourage this social aspect of walking, give residents choices, and distinguish your community from your competitors'. One option should be to allow people to walk briskly for a long distance as an aerobic exercise. Another can be a social walk past homes, shops, parks, and schools. These social paths should be enriched with places to stop, complete with benches, and maybe a water fountain. Another choice could be a nature path to be used to walk slowly, to look and listen and contemplate. None of these types of walks should be shared with bicycles. Bike paths can be close by or take entirely different routes.

David Jensen, a friend of Kephart's and a land planner of some renown, advises his clients to plan the open spaces and hierarchy of trails, paths, and walks before locating a single use or building on the site. Trails are too often an afterthought and are squeezed in anywhere, or they are located in drainage ways, too low for any view opportunities or for a proper sense of safety.

Other Amenities

Homebuyers also ranked indoor and outdoor pools high on their priority list in ProMatura's surveys. Responses varied significantly by income and geographical location, so it is vital that you survey your local market before making a decision on whether to include an indoor pool, outdoor pool, both, or neither. Nearly half of survey respondents who were buying homes in the $300,000 to $399,999 price range wanted a heated indoor pool, but only one-quarter of respondents buying homes in the $150,000 to $199,999 range considered it important. Nearly half of people with homes or prospective homes over $300,000 thought an unheated outdoor pool was important, but that figure went down as the house price decreased.

Survey respondents also expressed a preference for a neighborhood park, which is not something you can always control. Pet owners are plentiful everywhere and boomers are no different. Of the age 45+ homebuyers surveyed that were buying a home over $150,000, nearly one-third said they considered a dog park important. Again, regional responses varied.

Not all services that are important to homebuyers need to be within your development if they are available within a short bike ride or an invigorating walk. Boomers love to walk. Yoga studios, and wellness and fitness centers in the neighborhood can offer choices as opposed to having one tai chi instructor on staff who may not be qualified to guide people through a weight-lifting program.

Sidewalks and trails can be provided and connected to public extensions, even on the smallest of sites. At Rivers Bend in Cincinnati, Ohio, trails throughout the community lead to wilderness trails with blinds for bird watching and connect to public trails along the river. Walking as a major amenity has not been capitalized on to the fullest possible extent. Some communities have great nature trails, others have good exercise walking paths complete with signage detailing distances from point to point, and still others have created wonderful social walking environments. College campuses are the model for a social walking environment. There are benches for conversations; some are shaded, others are in the sun. Coffee shops and sandwich shops are along the way and everything loops on itself offering a variety of choices of ways to walk. However, they all fall short of offering everything in one comprehensive package in one community. Builders and developers are searching for ways to replace golf as the major community amenity because of the expense and the current oversupply of golf courses. "A Community on Foot" could be an answer. The Villages in Lady Lake, Florida, and Solivita in Kissimmee, Florida, both offer a golf cart transportation system of paths and roads to complement the streets and cars. A "community on foot" could take this idea one more step with a comprehensive walking plan to go along with the golf carts, further relegating cars to the garages for all but occasional trips outside the community. The buyers will consider this as an ecological design feature, first for not including a water-intensive golf course that leaches contaminants into the soil and secondly for reducing the dependence on the car and focusing on healthy walking instead.

Fitness

Fitness centers in 50+ communities get a lot of marketing hype and some of them actually get used. Usually they are better than the ones you find in small hotels but less well-equipped than you would find at a local gym or fitness center. Some of the largest communities can compete with the large chains but not many. You have the choice of either outfitting your fitness center and keeping the equipment in top shape (pun intended) or partnering with a facility close by. More gyms than ever are catering to the 50+ crowd and more are sure to follow.

"Nifty after Fifty" is a small independent gym in Whittier, California, specializing in fitness training for people over 50. It offers the full complement of weight training, cardio equipment, yoga and tai chi classes, physical therapy, dances, and even movie nights.

The purpose behind this "gray gym" is to encourage the unfit 50+ person to get on the road to good health, without the intimidation of seeing so many healthy 30-something bodies working out with them. Of course, 91-year-old weight lifter Willie Wortham, the fittest woman in the gym, could be a source of intimidation or encouragement.

Other programs and dedicated gyms for the 50+ abound today: Club 50 Fitness in Reno, Nevada, was the first of a chain of 50+ clubs across the country. Bally Fitness, along with many other clubs, offers "Silver Sneakers" fitness classes. Gold's Gym has 50 such clubs, as do 1,200 YMCAs.

David Hall, founder of LifeCenter Plus, has created a model fitness and health delivery system he believes can be replicated in 50+ communities. LifeCenter Plus in Hudson, Ohio, is a breeder facility that will offer training and consultation to builders and developers of 50+ communities. The LifeCenter Plus center is a 103,000-square-foot unique health and fitness center ranked in the top 100 in North America. It emphasizes fitness for older adults but is open to all families. Programs focus on well-being with a holistic approach. For the entire text of Hall's detailed description of LifeCenters, see the App. B.

Architecture for Homes

Functional design elements include space arrangements, circulation, closet sizes and locations, and entry placement. Site placement is part of the functional design, particularly placement for energy conservation and solar access. A long, rectangular house with the bulk of all its windows properly located and shaded in summer can save 30 to 40 percent on heating and cooling costs.

Heating and cooling systems in green/energy conserving design may have visible elements such as window shading, obvious orientation to the sun, solar panels, and high-performance glass in windows, all of which change the quality of visible light. Structural materials, insulation, and weatherproofing are invisible in finished homes but, nonetheless, many people care about them.

When age 50+ residents were studied, 55 percent of respondents said they worked with the builder to choose materials and only 1 percent of people buying homes over $250,000 did not care about structural materials. In another study, 77 percent of 55+ respondents were very concerned about the structural materials of their home and only 3 percent were not concerned at all.

Finish materials are extremely important to buyers and are the most often cited as evidence of the quality level of the home: 60 percent prefer either brick, stucco, or stone on the exterior; 65 percent prefer double pane glass; 77 percent prefer built-in shelving and cabinets; and 70 percent prefer stone kitchen countertops, according to a study by ProMatura a few years ago. No doubt quality-conscious boomers will deem finish materials even more important.

Kitchen design is a functional exercise that is often compromised by visual considerations. The emotional appeal of large islands has led to kitchens with multiple islands which aren't functional for a single cook or a couple. However, that spread-out work area may be just the thing for those who don't cook, but do hold frequent large gatherings where food is served by the staff of catering companies.

Visual elements of architecture are not only the things you see when you look at a home but what you see *from* the home. Finish materials, roof and building shapes, openings, and color and textures are some of the things you see when looking *at* a home. Views to the distance and in the foreground, outdoor rooms, gardens, the street in front, and light are things you see from a home. The materials, colors, etc., are also functional and an expression of the time and place a home is built. The design process sets priorities on all of these factors and the final design is a manifestation of the balancing of those priorities. Take views to the distance as an example. Homebuyers overwhelmingly choose homes with a view over those with no or mediocre views. When views are available due to topography and the placement of nearby homes, they may become a high priority to

Figure 8-3

A screened porch is a kind of out-door room, but with the warmth of the home inside creating a comfort-able feeling for visitors and residents alike.
(*Photo courtesy of KEPHART.*)

the designer and the future buyer. If those views are particularly spectacular they may shape the entire design of the homes and the neighborhood. One way to capitalize on beautiful views is to work to access those views from as many homes as possible in a development. Another way could be to locate a neighborhood park where the view is best and encourage everyone to enjoy the vista from "their own park."

Many homes are designed with little opportunity for views, even to their own backyard. It's an opportunity lost. Grouping windows and doors into one large glass wall facing the view is all it takes (see Fig. 8-3).

How a home feels is an experiential thing but, like views, parts of the kinesthetic experience are common among people. Raising the level of an adjacent patio to the floor level of the home can make that outdoor space feel like another room, increasing the perceived home size. Similarly, we know that a low light level is calming and relaxing, as is soft music; those elements make us open to the idea of romance and bring those feelings keenly into focus, while blocking out all thought of work or problems. Walking on deep carpet once felt luxurious; we still react to it, but now the carpet is just as likely to be an expensive throw rug in the conversation seating space only. Stone and masonry have a substantial appearance, giving an impression of sturdiness and longevity. Conversely, walls that sound hollow when rapped with the knuckles suggest that the home is "cheap and flimsy." Over half of 55+ homebuyers work with the builder in making decisions about structural materials, according to a 2002 study of the market.

Buyers may never reveal their deeper, more personal feelings to the developer or perhaps to themselves, and yet they may react to those their surroundings when shopping for a home for no apparent reason when they see or hear something. Being aware that these unconscious feelings exist is an important step to understanding your buyer.

Much of home design in America today places a high priority on history. Even the revolutionary new urbanism community designers have resorted to copying the past rather than designing homes that express today's technology, materials, and construction methods. The boomers may change all that.

Homes may be seen as possessions to be proud of or simply as functional places to rest and plan that next globe-trotting excursion to Asia or the Outback of Australia. Some friends of Kephart's have their house planned around the storage of fishing gear, hunting equipment, and their sports cars. They can park the appropriate vehicle in the garage and load it directly from their packing and storage rooms and they are off.

The surrogate family or families architects design for, we call "the market," a term that still connotes an impersonal connection to the process. Boomers want to have their say and they want the design response to be personal in some way.

Simple, unpretentious homes have a universal appeal. They are like a blank slate on which the homeowner can imprint their personality, free of the inflexible production housing process that inherently separates them from the design of the home. Such uncluttered designs are, furthermore, the easiest and most cost-effective to build using state-of-the-art production building techniques like panelization, factory-built modules and, most importantly, repetition of elements and groups of elements. Variety is achieved by creatively intermingling repetitious elements in various ways from home to home. Affordable communities result from this marriage of simple design and production, leaving the buyer opportunities to personalize their "blank slate."

Language can influence attitudes and can create unpredictable reactions. It is best to avoid the use of words like "the market" when referring to the people you are designing for. Shelve the term "product" when referring to homes for the same reason. It's a start toward finding ways to personalize design for homebuyers without paying the extra costs associated with a custom home.

Home and Community Synergy

Design features of the home and the community can work together to enable or enhance the experience of residents of a 50+ community. People who want to continue working or starting a new career and people who love to travel and want to do so worry-free depend on the synergy of their community and their individual homes.

New Careers

Even before the recession of the last couple of years, boomers said they wanted to either keep working, begin a new career, turn a hobby into a career, or spend time volunteering as their "job." A 2009 study by the Pew Research Center's Social & Demographic Trends Project found that over half of workers age 65+ work because they want to, and fewer than one in five cite needing the paycheck. Over one-fourth of respondents said they worked for a combination of the two reasons. Boomers have a greater desire to work than people 65+ and they will want their homes and communities to support them in that.

The demand for offices in homes is clear but it doesn't stop there. Community buildings or dedicated office buildings can provide space for community residents to rent to have an office outside the home but nearby. Apartment studios may be as popular as offices as boomers follow a long-held passion for artistic expression. Live-work housing forms can be a nice compromise between a home office or studio and one completely outside the home.

The lower level front room may be a studio or office with a separate entrance from outside. That space may be used for any purpose, including nonresidential; however, it is not required that the homeowner have a semicommercial function there.

These physical places in an active adult community or age-targeted community can facilitate the development of a new career for self-motivated people but more is needed for those who don't know how to get started. An on-site campus with staffing by a local university is being incorporated into many communities. Most are simple classroom buildings where after-hours adult education classes are offered. Village Homes' Observatory Village community in Ft. Collins, Colorado, has an observatory where college-level astronomy classes are held as part of a university program at Colorado State University. Other communities are more closely associated with a university, to the point that in some cases the active adult community is on university-owned land. For more on university-linked retirement communities, see Chap. 3 on neighborhood types. The idea here is to encourage you to support people in their pursuit of a new career via visiting lecturers, a public library connection, and the like.

To further support residents in their careers in an active adult community, where a much smaller percentage of homebuyers worked in previous generations, homes should have at least one office, possibly two if more than one person lives there. Not everyone desires or can afford an office or space outside their home so the space inside the home becomes even more critical for them to enjoy the experience and function well in their extended or new career.

Travelers

Nearly every research study on boomers mentions their desire to travel. Sixty-four percent of the 50 to 59 age group say that travel is their top unfulfilled ambition to take up in retirement, as reported by Del Webb in its 2005 Baby Boomer Report. This is a 20 percent increase from the 44 percent in its 2003 report.

Maintenance-free living including full lawn care and landscaping maintenance is not only a preferred amenity by more than 75 percent of 50 to 59-year-old boomers, it is the very thing travelers need to be able to leave home with peace of mind.

Frequent flyers and snowbirds alike would appreciate a packing area in or near their closets, complete with a convenient luggage storage area that doesn't require climbing up a rickety stair or ladder or crawling about in a hot, cramped storage space that harbors spiders and the occasional rodent (including squirrels). Basements are equally inconvenient, if not always as unpleasant.

Doggie day care and boarding of pets of all kinds is a valuable service to travelers. Other traveler services could include a part-time travel agent on site or accessible by phone and Internet, information on where foreign currencies are available, and where clinics are located that provide immunization shots and traveling advice. Outside travel agents could come in to show their latest tours that are available as part of the continuing education program.

Active adult community designs, with 75 percent of homes under 2,500 square feet, are great for travelers who may not have a second home and don't want to "rightsize" or clutter their lives with too many possessions.

Packaging all of these benefits for travelers into your Frequent Flyer Home offerings could be a marketing niche in itself. Model the packing area and the luggage and equipment storage conveniently located near the master closet and the garage to sell the idea. Equipment for travel such as fishing gear, backpacks, bicycles, scuba gear, skis and snowshoes, beach toys, etc., should be conveniently located to pack and leave.

REFERENCES

1. Del Webb, "Baby Boomer Report," Pulte Homes Inc., Bloomfield Hills, Mich, 2003 and 2005.
2. Taylor, Paul, Rakesh Kochhar, Rich Morin, Wendy Wang, Daniel Dockterman, Jennifer Medina, "Recession Turns a Graying Office Grayer," Pew Research Center, Washington, D.C., September 3, 2009.
3. Wylde, Margaret, *Boomers on the Horizon: Housing Preferences of the 55+ Market*, BuilderBooks (National Association of Home Builders), Washington, D.C., 2002.
4. Wylde, Margaret A., *Right House, Right Place, Right Time: Community and Lifestyle Preferences of the 45+ Housing Market*, BuilderBooks (National Association of Home Builders), Washington, D.C., 2008.

9 Design Options

"Design can be art. Design can be aesthetics.
Design is so simple, that's why it is so complicated."

—Paul Rand

As you put together your design room by room, the many options open to you may be overwhelming. Or you may not really know where to start. Or you might be looking for something different from what you have done before. Following is a guide of options and narrative from author Kephart, culled from years of his own experience and from tours of countless homes and home types throughout the world, plus input from author Schriener from many home tours and conversations with designers and builders over the past several years.

GREAT ROOM

The popularity of "great rooms" that encompass diverse areas—such as family living, nook, and kitchen—all in one well-defined space, has grown considerably in recent years.

Some see this popularity partially as a reflection of today's more informal lifestyle. Large nooks serve double duty as dining spaces, kitchens are wide open to the rest of the home, and often the living room can be dispensed with entirely to increase other room sizes without adding square footage to the house. Surely the flexibility of the great room space contributes to its popularity (see Fig. 9-1). Let's look at the spaces that could comprise a great room or be more defined, separate areas.

Living Room

Approximately one-third of buyers prefer a large family room and no living room, while two-thirds still prefer both, according to Margaret Wylde's study of housing preferences among age 55+ households in *Boomers on the Horizon*.

Figure 9-1

A great room includes living, dining, kitchen, entry, and even a den in this case. (*Drawing courtesy of KEPHART.*)

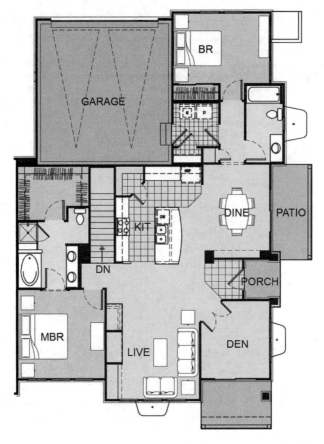

Townhomes - Ranch
1,544 sq.ft.

Living spaces, as in the case of dining rooms or parlors, make good flex areas for offices or additional bedrooms. Kephart's experience differs from the research on preferences. In communities he has designed, the wide-open plan that has fewer defined rooms and has a great room is invariably the most popular.

Dining Room

Seventy-one percent of 55+ buyers want a dining room. Most of those, 60 percent, want the dining room to be open to the living areas and 40 percent want it more strongly separated with walls or some other divider.

With the plan illustrated (see Fig. 9-2), a builder is able to satisfy the larger group of buyers. But what about the 29 percent of buyers who don't want a formal dining room at all? A plan option that offers both choices can be a solution for everyone.

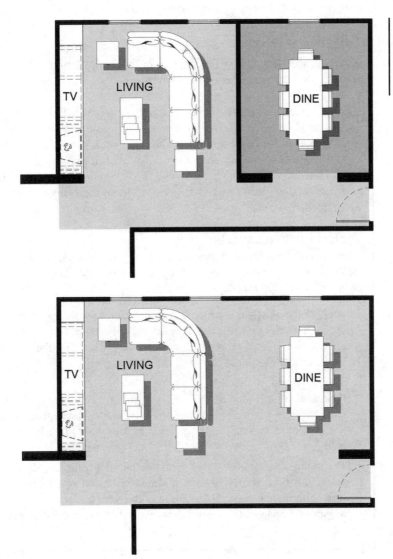

Figure 9-2
Options for dining
and living rooms.
(*Drawing courtesy
of KEPHART.*)

People who are comfortable using a large nook or informal eating space in the family area can use the dining space for an office, two offices or get-away rooms, an art studio, or any other use they can imagine. The dining space can even become a guest room if it is properly located with access to a bathroom. The area could also simply become additional living space.

Kitchen

Two-thirds of 55+ homebuyers want a completely open kitchen or one separated by a half-wall from the family room, according to Wylde.

Kitchen Wisdom

Kay Green of Kay Green Design Inc. in Orlando, Florida, offers her advice on what to consider when designing a kitchen.

"Would a freezer on the bottom instead of the top or side be helpful? Would a cook stove with knee space below work better than a conventional oven with burners on top? Where will it be positioned in the kitchen? Is a microwave oven or toaster oven used more often than a conventional oven? If it isn't being used, lose it. What about knee space under the counters and sink? A rolling cart helps in moving dishes and pots around easily, so you may opt to leave more floor area, if possible, to accommodate a cart.

Typical Kitchen Layouts

These kitchen layouts work best for cooks in wheelchairs, or cooks who have difficulty in walking or standing for long periods of time. The layouts provide users with the most efficient use of space.

- An L-shaped kitchen is perhaps the most user-friendly, especially if the kitchen space flows right into the dining area or family room. You'll get maximum floor area with the least amount of walking or rolling. Appliances should not be located too close to the corner of the "L."
- The U-shaped kitchen has two parallel walls connected by one short one. In the middle of a U-shaped kitchen, everything is conveniently located on either side, but here again, appliances should be located away from the corners and, if the kitchen is small, no appliance or sink should be located in the short wall.
- The galley kitchen is designed with two parallel walls, similar to the U-shaped kitchen. A third wall, if any, is short and serves only to connect the other two. Providing there is enough free space in the middle—at least a 60-inch diagonal turnaround space for wheelchairs—everything is reachable from the left or right.

It has been Kephart's experience that homes that incorporate the kitchen as part of the living space are usually the most popular choice with buyers, versus a plan with both a living and dining room divorced from the kitchen.

Certain builders have taken the idea of the completely open kitchen to new heights. In one model, the kitchen is largely one long wall, with a huge island, opening to the informal eating area and family room, all contained within one large space. This home has a separate dining room but others simply lengthen the great room and place the dining table immediately adjacent to the kitchen island (see Fig. 9-3).

The truth is, each of the dining rooms in the previous examples could easily be used as an office or an additional bedroom, because the large informal table in the family room is so

Figure 9-3
Red oak kitchen, no barriers between dining and kitchen. (*Photo courtesy of Kay Green.*)

pleasant and close to the action in the kitchen. On plans, that space is labeled as a dining room because, according to Wylde, 71 percent of 55+ buyers still want a formal dining room.

A completely open kitchen concept is enhanced when the breakfast bar is on the same level as the kitchen counter tops, allowing the free flow of space. A simple one-level kitchen island makes a small home feel more spacious (see Fig. 9-4). It is a good idea to keep the countertop on an island or a peninsula at one level; it can still be used as a breakfast bar with shorter stools.

Still, 16 percent of 55+ buyers said they want a completely separated kitchen, according to Wylde. Sometimes this is a cultural thing. Kitchens in Turkey and India, for example, are as far away from the living area as possible, and people from those countries and others are uncomfortable with the informal lifestyle so prevalent in the United States, including our wide-open kitchens.

Kephart and his wife Jaye stayed in an architect friend's apartment in Istanbul. Notice in the plan shown in Fig. 9-5 that the kitchen is away from the large living and dining area. While working as an architect in Turkey, her attempts to open up kitchens and living areas to each other in production homes for sale have met with limited success. Some people ask for a wall with a door to the kitchen.

In India, some homes are designed in accordance with the Hindu spiritual design practice called *Vastu Shastra*. According to this practice, the kitchen must be in the southeast corner of the house or apartment.

Figure 9-4

One-level counter-top and island makes the home feel more spacious. (*Photo courtesy of KEPHART and Steve Hinds.*)

Figure 9-5

Plan of apartment in Istanbul, where kitchen is closed off and entrance is far away from living/dining room. (*Drawing by Mike Kephart.*)

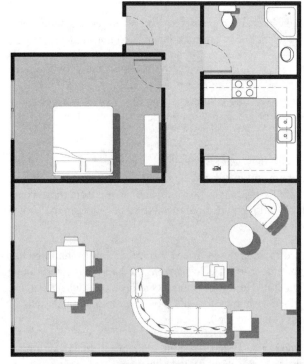

ESAN'S APARTMENT
IN ISTANBUL

BEDROOMS

Fifty-three percent of all 55+ buyers want three bedrooms, but older groups are most likely to prefer two bedrooms, according to Wylde's research. Forty-seven percent do not want a three-bedroom home.

Not many people want four bedrooms or more, and so a sizable percentage of 55+ buyers want two or three bedrooms. Part of the problem is the labeling of rooms. Buyers want offices, get-away private places, TV rooms, art studios, etc. It's quite conceivable that a couple could want only one bedroom for themselves and use extra bedrooms for other purposes, or look for a large one-bedroom home with oversized living spaces for entertaining. There's always a trade-off when it comes to space. But we Americans do like our bedrooms; 68 percent of respondents to Wylde's study said they would prefer a large master bedroom to a larger bath.

When Kephart and his wife have their home appraised, they label all those extra spaces as bedrooms, but in reality they have one bedroom, an exercise studio, a meditation room, and an office for the two of them. The meditation room does double duty as a guest room, and the exercise room is also Kephart's closet and dressing room (the wall of mirrors serves both functions). The office is for that use only. They have an eating space, which they call the dining room, but it also serves as a place to fold laundry and, with full-height bookcases, it is their home library. Flexible design is certainly one way to allow a single floor plan to serve many different needs. The labeling is meaningless; it's the use that matters.

Seventy-one percent of the 55+ buyers in Wylde's study want a dining room and 41 percent want a home office. The option for either use will go a long way with buyers. Why can't a bedroom with walls removed become a dining room with a small, nonstructural change?

Some three-bedroom homes split the bedrooms with the master on one side of the living spaces and bedrooms two and three on the opposite side, giving the master suite as much privacy as possible (see Fig. 9-6). Or, the bedrooms may be split while on the same side of the home. This arrangement has proved the most popular choice among nonfamily buyers for years and in some cases also offers the opportunity for flexible use of the extra bedroom spaces without involving revision of the master suite. One desirable option that cannot be offered in this type of plan is the conversion of an extra bedroom into a master sitting room or office connected to the master bedroom.

Plans that group the three bedrooms are less popular in general, but they do offer the chance to create a second sleeping space for the snoring partner, which is not something people readily reveal to a salesperson. The grouping of three bedrooms down a single hall makes it more difficult to provide flexible uses that more closely relate to the living spaces. There are many exceptions to these generalities but it's important to keep flexible use possibilities in mind when designing homes for boomers. If you are offering several new home plans in your community, it would be good to split the

Figure 9-6

Master suite separated from other bedrooms is most popular choice of boomers. (*Drawing courtesy of KEPHART.*)

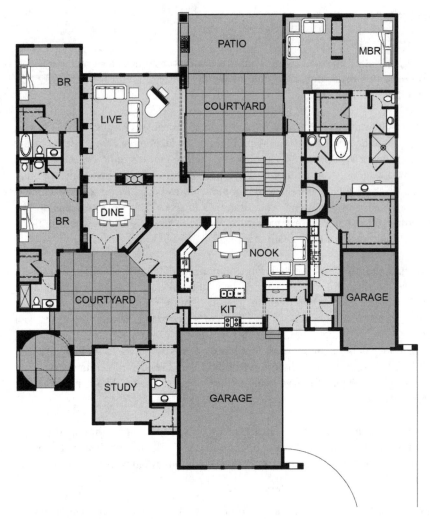

bedrooms in most of those models to offer the things buyers want: privacy for the master suite and flexible use opportunities for the secondary rooms. An additional model or two can have at least one additional bedroom space close to the master suite as a separate sleeping room or as a nursery for those grandparents taking care of infants or young children.

MASTER BATHROOM

Bathrooms have gone through significant design changes over the past several years. One choice to make is where to have the closet in relation to the bathroom. Should it be between the bedroom and the bath? Or should homeowners have to go through the bathroom to get to an often cavernous walk-in closet?

One of the most debated design challenges in bathrooms has to do with the toilet. Seventy-eight percent of 55+ buyers want a separate shower and a private toilet compartment, Wylde says.

Bath Wisdom

Kay Green of Kay Green Design offers these suggestions for bathrooms:

- *Safety in the bath:* Provide a walk-in bathtub. Benefits to disabled persons include a walk-in door (no stepping on the edge of the tub), water-tight door, high seats (no lifting or lowering body weight), and knobs that allow for easy water temperature adjustment.
- *Folding tub seat/transfer bench:* This is a seat that folds down to sit at the level of the tub edge. You would swing your legs into the tub. It can also be used in the shower by mounting to the shower wall.
- *Sink accessibility:* Remove vanity cabinets and replace with a wall-mounted lavatory, since pedestal-style sinks are usually too tall. Insulate hot water pipes under the sink to prevent burning.
- *Bathroom door:* Width should be at least 32 inches for a wheelchair to fit through straight on. If entering at an angle, the doorframe should be 36 inches wide to allow the wheelchair to fit through.
- *Grab bars:* Used by the toilet, tub, or shower; should support 250 lb, mounted horizontally or vertically, depending on the space.

Builders need to come to grips with a contradiction. A private toilet compartment is not accessible or easy to use unless it is much larger than the typical ones offered on the market. Even if there is not a grab bar in the private toilet compartment, there should be room for a small cabinet, hamper, or something that the owner or guest can use to aid in rising from the toilet if necessary.

In a more recent study by Margaret Wylde, 57 percent of the 55 to 64 age group said that a more accessible home is one of their major reasons for moving. Twenty-two percent said it was their primary reason.

The separate shower, however, is a step in the right direction for ease of use and accessibility. The last step, literally the one at the door to the shower, needs to be eliminated to do the job properly, as shown in Fig. 9-7, although wheelchair users generally can navigate a short edge into the shower.

Twenty-two percent of the 60 percent of age 55+ homeowners expected to move in the next 15 years is over 6 million people who want an easier to use and more accessible home. Quite a market niche, isn't it? Include the 57 percent who listed accessibility as one of their reasons for moving and the size of your potential market triples.

Figure 9-7
Roll-in shower.
(*Photo by Mike Kephart.*)

OFFICE

Thirty-nine percent of the 55+ have an office in their home, according to Wylde's study. Forty-four percent of singles have offices and over 50 percent of those with incomes over $50,000 per year have offices. Forty-three percent of respondents in Del Webb's 2005 Baby Boomer Report, said they will continue to work in retirement and 44 percent were not sure whether they would work or not.

Home offices may be designed into homes in a number of ways to respond to the various needs for an office. Some buyers may want a formal business office designed to receive business guests; others may simply need a small place to quietly work. Some may want a shared space and others may need seclusion to read, meditate, or work.

Not all offices have to be inside the house. The separate "casita," as it's called in the Southwest, or the "summer kitchen," as it is known on the shores of Lake Erie, is a faint reminder of the history of the area. The separate house idea originated in times when household servants were common and these buildings either housed the servants or, in the example of the summer kitchen, served as the kitchen during the summer to keep the heat and odors of cooking outside the home. This cooking arrangement can be found in historical homes around the world but it vanished when cooking duty was shifted to the woman of the home as the servant class became independent.

Today, these casitas, summer houses, guest houses, pool cabanas, or pavilions can make wonderful offices for those looking for seclusion and privacy while working.

A room above a garage with a separate entry that doesn't go through the house can serve a similar purpose. These granny flats or guest houses are often found in traditional neighborhood design (TND) communities as rental apartments or to house an adult, child, or relative. And, of course, the simple bedroom/flex office space has been with us for a long time.

Offices don't always need to be entire rooms or use that much space. A corner of the family room can serve as an office or a small space off the kitchen nook can house an office in half the space of a bedroom. An extra wide corridor can accommodate a desk and workspace quite nicely, as can any transition or circulation space.

One fairly common layout is to create a library/office leading from the foyer to the living room. In this case the office is up front and serves as a greeting space when welcoming visitors.

Offices can be part of almost any room or an alcove off of any room. They can even be concealed behind cabinet doors to shut away the clutter when entertaining guests.

STORAGE

According to Wylde's *Boomers on the Horizon*, 63 percent of 55+ buyers overall don't believe today's homes have adequate general storage.

Inadequate storage space or the lack of it altogether is a common source of complaints from homebuyers. There are several ways to address this concern: basements, larger garages, attic storage, or planned additions between house and garage.

Nearly two-thirds of respondents in Wylde's study wanted a half or full basement. If a basement is included for the purpose of additional living space and/or storage, the stair should be located and oriented in a way that facilitates moving items up and down the stairs. Stairs that turn or are reached by a narrow hallway are difficult to use for that purpose.

If the basement is a "walkout," be sure the stair terminates at the lower level facing the available natural light source, that is, windows or sliding doors.

About half of all age 55+ couples want a two-car garage. About 21 percent of the 55- to 64-year-old boomers want a three-car garage; 11 percent want a one-car garage; and 2 percent want a four-car garage. Thirty-seven percent of those buying homes costing over $250,000 want a three-car garage.

It costs very little more to oversize a garage space enough to add general storage. The garage can be wider to add space on one or both sides (his and hers) or it can be extended in depth to add storage in the front of parked cars. Be sure to give people room to access shelving without having to move the cars.

Another way to provide accessible storage in garage attics is to extend the depth of the garage by 4 feet and to include a permanent stair parallel to the house wall. Pull-down stairs are an option but require moving cars and an athletic person to climb the steep ladder.

Adding space for general storage between the house and garage comes at more of a price than the previous two options, but it can be optioned as finished space for those who choose to add an office, morning room, larger kitchen and nook, or a larger laundry or hobby space. The morning room option was modeled in the Arbors sales park in Brunswick Hills, Ohio, and has proved to be the most popular—the usual reaction to modeled homes and options.

LESSONS LEARNED FROM EXPERIENCE

Much of what Kephart learned about homebuyers was validated by the research done by Margaret Wylde, AARP, and Del Webb, but some things are difficult to ferret out in formal research. The examples that follow demonstrate things he has learned directly from homebuyers.

Give Them Sleep Options

Many couples sleep apart some or most of the time. They are still a loving couple, but one snores or is restless in sleep, which disturbs the other. They both may still use the master closets and the master bath together, or one of them, usually the man, uses a second bedroom, its closet, and the second bath for convenience. They may also use the second bedroom when one of them is ill. Making it easy for people to make these decisions by providing simple options is one way to address this need without intruding in the couple's personal life. Locate a second bedroom close to the master and option it as a sitting room with additional access options to allow multiple access points into the master bath and closets. Provide a large walk-in closet in the second bedroom for the person who may use it instead of the master closet, and locate a second bath close by to give them additional use options.

Give Them a View upon Entering

Kephart has seen, time after time, that homebuyers' overwhelming favorite model homes have "see-through" plans. The moment you open the front door you feel a sense of spaciousness. Every living space is open to the others with minimal definition of individual rooms by walls, floor material changes, or ceiling heights. Your eye is irresistibly drawn to the view outdoors in the rear of the home and the abundant light flooding through the large glass area that, ever so slightly, marks the line between indoors and outdoors. The flooring material often flows to the large patio and courtyards beyond with no level change. The impression is created that the living space continues without limit. These buyers are not concerned about having to negotiate their way through the foyer, dining room, kitchen, and family room without a clear circulation path to get to that rear glass. It's the view that's important to them.

While it's true that the home design described above is the first to sell and will usually outsell all other models, it isn't for everyone or every location in the community. Such an open plan depends a great deal on what the rear view may be. Homes that have rear-loaded garages, for instance, cannot have that wonderful relationship with the outdoors, at least not to golf courses or open spaces.

Not every home-site in a community can have open space at the rear. In fact, the overwhelming majority of home-sites in most community plans have small backyards and the view is of the neighboring home's rear elevation.

When lots are small in order to be affordable, these yards grow smaller and the view can become more negative than positive. These conditions demand more creativity in providing the necessary outdoor space without the loss of privacy. Courtyards are particularly effective here. They can become "outdoor rooms" and the courtyard itself, its fountains and furniture, become the view.

Consider a Courtyard

A courtyard provides a terminus for the eye past the glass walls by locating interesting elements such as fireplaces or fountains as far away from the outside wall as possible. Courtyard homes turn more inward than outward because of the closeness of neighboring houses. Taken to the ultimate, a courtyard can become a selling point for a home.

In Fig. 9-8, everything revolves around the beautiful courtyard living area. Note the absence of a level change from kitchen to courtyard. Not only does this detail make the courtyard more accessible, it also creates a stronger impression that the inside and outside rooms are all part of the overall design.

Put the Master Bedroom Suite on the First Floor

Boomer buyers will not hesitate to buy two-story homes as long as the master bedroom suite is on the same level with the kitchen and living spaces, but the circumstances matter. Community locations close to the action—with shopping, recreation, and services—are in great demand, but achieving a relative level of affordability requires a trade-off. In that circumstance, homes with main floor masters are quite popular. The research is relatively silent on this home configuration choice, or the choice of a two-story with elevator home design.

The same is true of attached homes such as townhomes. These housing forms are a good choice when a first-floor master bedroom is included. Purchasers get one-level living for their daily lives and a more affordable home when compared to a ranch or single-story plan. Two-story model homes and attached homes are seldom well received when one-story/ranch homes are offered in the same location, unless the one-story homes are significantly more costly.

Figure 9-8

Courtyard view looking out. (*Photo of One Cherry Lane courtesy of KEPHART and Steve Hinds.*)

Builders can create their own undesirable competition for two-story, main-floor master homes if they price the ranch homes too low or offer too many additional space options in the other plans. Then they can sit by and watch all the ranches sell first and struggle to sell the remaining two-story models.

Kephart has watched as experienced builders have "tested" a two-story, main-floor master in their predominantly one-story active adult community. First of all, the two-story was not modeled in the model park, as all the ranches were, but built as an experiment and shown only to buyers willing to make the trip to the home's remote location. The poor location and the lack of marketing can make any plan offering fail, as most builders will acknowledge. Two strikes!

The third strike is the "not quite perfect" two-story configuration in a sea of one-story homes. The "test" predictably fails, and it will fail every time.

REFERENCES

1. Del Webb, "Baby Boomer Report," Pulte Homes Inc., Bloomfield Hills, Mich., 2003, 2005.
2. Wylde, Margaret, *Boomers on the Horizon: Housing Preferences of the 55+ Market*, BuilderBooks (National Association of Home Builders), Washington, D.C., 2002.
3. Wylde, Margaret A., "Active Adult Housing Prospects," National Association of Home Builders Webinar, September 2009.
4. Wylde, Margaret A., *Right House, Right Place, Right Time: Community and Lifestyle Preferences of the 45+ Housing Market*, BuilderBooks (National Association of Home Builders), Washington, D.C., 2008.

10 Looking Ahead

"I like the dreams of the future better than the history of the past."

—Thomas Jefferson

TRENDS

Here a Trend, There a Trend, Everywhere a Trend, Trend

Looking to the future and naming "trends" is a popular pastime at homebuilder conventions and award ceremonies. The authors attended every presentation on the future of home and community design until they were dizzy with prophesies, most of which were exaggerated and few of which ever came true. After a time Kephart was asked to join the speakers on stage to offer his insight. It seems that after you age awhile and continue coming to these presentations, it looks as if you might have something to share. He eagerly joined in and thoroughly enjoyed the exercise in the beginning. Speakers attempted to identify trends they saw each year when judging competitions or just observed in their businesses.

The problem developed for Kephart when each year they had to come up with the top 10 trends, but things didn't normally change that rapidly from year to year. The ad hoc swamis stretched themselves and one year Kephart remembers putting his stamp of approval on the most obscure trends imaginable. He doesn't have them at hand and wouldn't embarrass those with him in the effort, but a typical trend presentation went like this:

Trend: "The designs this year displayed a newfound sensitivity in the expression of materials texture and detail."

Kephart's thoughts as the list went on in similar fashion: "HUH? What the heck happened to simple prevailing concepts such as, 'The home designs were less complicated this year, showing that the emphasis on community design continues to prevail'?"

That experience was Kephart's breakthrough into the full realization that it isn't the big change each year that is the most profound, but the one that continues to grow in importance over the years. Schriener, as a journalist who for years had been witnessing similar trends presentations at conferences for every aspect of the design and construction industry, also had figured that out. One of her favorite books has been *The Black Swan*, in which author Nassim Nicholas Taleb contends that some of the occurrences that have the biggest impact on us are ones we can't predict—the ones that blindside us—and we struggle to come up with a more or less rational explanation for each one after the fact so they don't seem so random and unpredictable. The events on September 11, 2001, and the amazing success of Google are two examples he gives in the book. So you've got two authors who are a bit jaded about predictions trying to come up with something of value to share with their readers in a chapter about predictions. Well, they gave it a shot. Read on.

The recession was the big news the past couple of years and could be called some kind of a trend, but what happened in the past 2 years wasn't the trend; it was what was done over the last two decades that finally surfaced. Just as with 9/11 and Google, it wasn't that nobody knew what was about to happen. It was that either few people were in a position to see what led up to it, or the rest of us just weren't paying attention, or both. Taleb says to focus on the generalities and not the specifics if you don't want to miss opportunities. Some people saw the recession coming but most of us missed the muffled, muted signals from the early warning system and got whacked.

Kephart stopped naming trends and let others take that job since it continues to be one of the most fun party games at industry conferences each year. Schriener kept attending the trends sessions at conferences in hopes of increasing her understanding of the industry she covers and out of fear of missing something if she didn't. Every year the interiors people talk about color trends, furniture styles, and Eastern or Polynesian influences. The appliance manufacturers tout their latest in-drawer refrigerators and microwaves, plus the new colors coming out. It seems that every discipline within the larger group of suppliers to homebuilders has its own "top 10 trends" for the year, but in doing so they sometimes find they have to scrape the bottom of the barrel to find anything that looks like a trend. Architects do the same thing. For instance, does anyone really believe that earth houses are really a trend that is growing, or that the Eastern spiritual practices of *Vastu Shastra* or feng shui will someday transform housing design?

Kephart often takes a fallback position on trends now and instead relies on the major forces of change, but that's no fun. It takes years and years before people notice they are growing older, becoming their fathers and mothers right there in the mirror—but nowhere else, thank heavens. It's the same with the slow-moving hands on a clock. It's hard to see the movement when you're so close to it. So, maybe the authors will take a step back, look at the long view, and play a little bit here. Since neither author is an economist—Kephart is an architect, Schriener a journalist—they will steer clear of the economic trends. There are enough others out there playing that part of the game. The authors promise not to use any "feeling" words such as breathtaking, warm, loving, or cozy. They will try to be as factual as possible when describing a thing as ethereal as design.

In no particular order, the authors offer the following observations:

Trend #1: Homes Are Getting Smaller

Though they didn't quantify how much home sizes have shrunk, the American Institute of Architects reported in their Home Design Trends Survey for the first quarter of 2008 that sizes of both custom luxury homes and entry level homes, have gone down. The U.S. Census Bureau, however, did report real numbers. According to a Census Bureau report published in 2009, in just one-half of the year, homes newly under construction are, on average, 300 square feet smaller. Jim Phelps, a builder of large custom homes the authors interviewed noted that the market for homes in the 5,000-square-foot range is now down to 3,500 square feet or even 3,000 square feet. His 10,000-square-foot buyers have recently moved down in size as well. He cited a concern for the environment as the major influence on this group of homebuyers who are less impacted by the economic conditions. KB Homes CEO Jeffrey Mezger noted a reduction in his company's median home size of 200 square feet as early as 2007. The current market conditions, the concern for the environment, and the impact of older singles and couples entering the market as they age all contribute to the attraction of smaller homes.

Trend #2: Three-Generation Families Are a Rapidly Growing Segment of the Market

USA Today's Sharon Jayson wrote in August 2008 that this market segment grew by 38 percent in the 1990s but was still a small segment of the total market for homes. In 2000, according to the U.S. Census Bureau, 2.3 million households included older parents living with their adult children; in 2007, that number had jumped to 3.6 million, a 55 percent increase. Neither number includes other relatives living in the households, which in 2007 totaled 6.77 million. The recession, which occurred after those surveys were conducted, encouraged even more families to move in together to share one home versus many. Multigenerational households is a significant market niche that only a few are paying much attention to, but this market, like all others, will someday be "discovered" by homebuilders.

Trend # 3: Outdoor "Rooms" Are Growing as Home Sizes Decrease

It only makes sense, with home sizes shrinking, that architects and designers would begin to utilize the outdoors as an integral part of the living space. We've always had patios but these were always thought of as "outdoors." A patio was down a few steps, made of ordinary gray concrete, and if furnished at all it was with functional plastic patio furniture. The trick is to switch that thinking to "outdoor rooms." Furnish these rooms as completely as any other room, tuck the outdoor room into the form of the house for protection from wind, and locate them to gather the warmth of the sun, or in the shade, depending on where in the country you may be. Last, and maybe most important, bring the patio floor up to the exact same level as all the rooms inside the home. It feels more like a "room" when that is done. Use the same material for flooring if possible—the flow of space is more seamless—and use floor-to-ceiling glass between the indoor and outdoor rooms for the same reason. Sliding glass doors will do when constrained by budget, but folding or pocket sliding glass panels do the best job. You can see an example on the cover photo of this book.

Trend #4: Green Is Here to Stay

This may be stating the obvious, but it is clearly one of the major trends of this and the last decade. As stated earlier, cities, counties, and state governments are racing to legislate green practices in spite of a healthy volunteer program instituted by the National Association of Home Builders in 2004.

Some of the green/sustainability programs in place today are

- NAHB's Model Green Homebuilding Guidelines.
- LEED (Leadership in Energy and Environmental Design) by the U.S. Green Building Council (USGBC).
- Energy Star, a program of the Environmental Protection Agency, is largely an energy rating system for equipment and appliances, but it is expanding into other green programs.
- Built Green Colorado and many other local programs in Atlanta, Chicago, California, and many other places that have chosen to adopt their own standards rather than use one of the top three national models. Over time this will lead to confusion and unnecessary complications for companies that work in several jurisdictions.

Today it looks like the LEED program, which is being used by a majority of jurisdictions, may prove to be the standard for building green in the future, but none of this matters nearly as much as what has happened to public opinion in the last few years. This book cites several of the ProMatura Group's surveys on green building and how much people are willing to pay up-front for energy efficiency. With the exception of 2009, the number of consumers that said they would pay to save $1,000 per year has been growing. The recession may be the reason for this change and, if so, probably as soon as the market improves energy efficiency will rise to the top of the priority list again. Meanwhile the auto industry is reinventing itself with a greater priority on fuel efficiency, people choose products to buy at the supermarket based on their green qualities, and peer pressure to be green is increasingly evident everywhere. (Remember that neighbor of Kephart's with the Prius?)

Trend #5: Accessory Dwelling Units (ADUs) Are Finding Encouragement from Cities

With more families taking in adult children, grandchildren, or parents, the demand for small second dwellings on single-family home lots along with the primary home is sure to increase. Especially in the case of aging parents, it makes economic sense, as the cost of living in an assisted living facility could range from $24,000–48,000 or more per year. It seems that nearly every major city in the country is considering joining the crowd and adopting legislation to allow ADUs (accessory dwelling units). In 2008, HUD released a case study on ADUs and the following is from that report:

> In response to suburban sprawl, increased traffic congestion, restrictive zoning, and the affordable housing shortage, community leaders began advocating a change from the sprawling development pattern of suburban design to a more traditional style of planning. Urban design movements, such as Smart Growth and New Urbanism, emerged in the 1990s to limit automobile dependency and improve the quality of life by creating

inclusive communities that provide a wide range of housing choices. Both design theories focus on reforming planning practices to create housing development that is high density, transit-oriented, mixed-use, and mixed-income through redevelopment and infill efforts.

In the late 1970s to the 1990s, some municipalities adopted ADU programs to permit the use and construction of accessory units. Many of these programs were not very successful, as they lacked flexibility and scope.

Although a number of communities still restrict development of accessory dwelling units, there is a growing awareness and acceptance of ADUs as an inexpensive way to increase the affordable housing supply and address illegal units already in existence.

AARP released its Model Code for ADUs in 2008 as well. The following is an introduction from that model ordinance:

> Accessory dwelling units (ADUs) are independent housing units created within single-family homes or on their lots. These units can be a valuable addition to a community's housing stock. ADUs have the potential to assist older homeowners in maintaining their independence by providing additional income to offset property taxes and the costs of home maintenance and repair. Zoning ordinances that prohibit ADUs or make it extremely difficult for homeowners to create them are the principal obstacle to the wide availability of this housing option. The model state act and local ordinance are intended to assist AARP volunteer leaders and other interested citizens, planners, and government officials in evaluating potential changes to state laws and local zoning ordinances to encourage the wider availability of ADUs. This legislation has been drafted to meet the needs of a wide variety of communities. Optional provisions, including those attractive even to very cautious communities, are incorporated in the model local zoning ordinance to provide as many choices as possible for jurisdictions to consider.

The Denver Council of Regional Governments (DRCOG) has instituted a study on ADUs, to be followed by its recommendation to all the municipalities in the Denver metropolitan area. Already more than a dozen cities have adopted their own regulations allowing ADUs, and more are sure to follow. Even Denver is including ADUs in the rewrite of its new zoning code to be adopted by city council in early 2010.

Trend #6: Technology Is Getting Smarter but We Aren't Necessarily Following Suit

Technology already *can* do so many more things than we use it for. By remote control through a cell phone, you can turn the lights on in your mom's living room 2,500 miles away, change the channel on her TV whether she's watching it or not, monitor your vacation home after a heavy rain to see if it flooded in that corner you're always worried about, turn your alarm system off to allow in the neighbor who's watching your house while you're gone, then turn it back on, and adjust the room temperature so it's just right when you get home. It's not just *Star Wars* fantasy stuff or "vaporware," as the techies used to call software that sounded great in the description but never materialized. It's real. We can do this now. Ah, but do we do all of those things? Any of them? Probably not.

It's not necessarily that new technologies are too expensive. If they do what we want and need for them to do, they are probably worth it. "Need" is a relative term. In 1992 when author Schriener got her first cell phone, she was very nervous when she used it while walking on the streets of New York City where she lived and worked, fearing that someone might see the phone and steal it. Cell phones were a luxury that a small minority of people had back then and it was considered quite rude to talk on one when in the company of others. Now who can live without their cell phone? It has gone from a luxury to a necessity and the more functions we can perform on it, the more necessary it will become. Schriener's global positioning system (GPS) is another example. Many people would consider a GPS in a car a luxury. But when Schriener moved to Columbia, Maryland, one of the most convoluted street systems in the universe (like where Little Patuxent Parkway crosses Little Patuxent Parkway), she literally couldn't find her way home at times without it.

One of these days, we will all have the set-up described at the beginning of this "trend." But for now, we don't, and that's probably less because we can't afford it and more because we fear trying to figure out how to set it up, put in the information it needs to make it control all of those devices, get the settings right, use it correctly, and get out of any trouble we may get into while using it. Manufacturers are trying to make technology as simple to use as they can. But consider that even now, not everyone even has a cell phone and some still use it only for emergencies. Who knows how many people can't figure out how to retrieve the messages in their voice mail. It's a big leap from that mentality to controlling a home's myriad functions by remote control. So technology can race on, but it's getting more and more difficult to keep up with it. Just as the gap is widening between rich and poor, the gap is increasing between the tech savvies and the tech bumpkins.

Trend #7: Boomers Are Coming up with New Ways of Using Their Homes

It used to be that people lived in their homes or rented them out or left them empty while they went on vacation. Those were the options. Then time shares came along, giving people the opportunity to have a homier atmosphere and pay less when they went on vacation to a nice place. Boomers don't like being tied down to a certain week each year or even a certain place. Time-share-swapping Web sites popped up and that solved some problems. One friend of Schriener's traded a week in her condo in Maui for a week at a stranger's in Turkey.

Now people are getting more creative about how they are using their homes. Boomertising CEO Priscilla Wallace sees homeowners treating homes like they do cars in the not-too-distant future. Lease one for a few years and then turn it back in and lease a new one or buy it if you choose to. That plays to boomers' penchant for changing their lives every few years, flowing through life phases, and keeping their options open. Other creative ways of dealing with home occupancy are sure to blossom as boomers indulge their desire and perceived God-given privilege to travel (see Fig. 10-1), Wallace thinks, once everyone sees that they're not limited to old ways. A house can be treated like a boat or a car or skis, not necessarily just like an immovable object that keeps boomers tied to it as well. Once again, boomers will lead the way.

Figure 10-1
The coast of
Turkey, one of
the new tourist
areas Americans
are spending their
dollars to see.
(*Photo by Mike
Kephart.*)

Trend #8: The Lines Are Blurring between Product and Service Sectors

It used to be that active adult communities had no use for the likes of affiliations with healthcare facilities (see Fig. 10-2), let alone continuing care retirement centers (CCRCs). The active adults didn't want to see old, frail, sick people or be reminded that it could be them one day. Never the twain met. CCRCs had their territory, active adult communities theirs. Now, however, you're starting to see CCRCs expanding their offerings to include larger, more apartment-like units for their residents who need assisted living services; and some are not only doubling the number of independent living units they offer, they're also doubling the size.

University Place, which is affiliated with Purdue University in Indiana, was built in 2003 but already is expanding to respond to updated demands for more units and more space. Independent living apartments that started out in 2003 to be a minimum of 600 square feet are doubling in size to a minimum of 1,200 square feet. Another CCRC, Holy Cross Village at the University of Notre Dame, has expanded and added units of up to 1,600 square feet versus the older units that were only about 700 square feet.

Maybe most surprising of all is the direction more active adult communities are taking. The trend is for them to look at adding a healthcare component or affiliating with a healthcare facility nearby.

Certainly the demand will be there. As always with boomers, the emphasis will have to be on health, not on illness or deterioration. Currently there is just one physician specializing in health issues of older people for every 8,500 boomers, claims celebrity

Figure 10-2

Assisted Living, the community type on the horizon for boomers, or maybe something else. (*Photo courtesy of DTJ Architects and Planners.*)

Martha Stewart in her December 1, 2008, blog posting on The Huffington Post. She donated $5 million to New York City's Mt. Sinai Hospital to establish the Martha Stewart Center for Living, which opened in 2007. It houses physicians of all types, social workers, nutritionists, classes and lectures for exercise, meditation, and caregiving. Stewart describes it as a "facility for the coordinated, comprehensive, outpatient care of older adults leading more robust and vital lives. It is not about aging. It is about living–living gracefully and healthfully with energy and enthusiasm even as we grow older." With baby boomers at the helm, we can look forward to more health-oriented integrated services affiliated with healthcare facilities in the next decade and beyond. That will make them more attractive to boomers and the owners and managers of the communities in which they will live. Again, these systemic improvements will be part of the boomers' legacy for following generations.

In a few years, it could be that both types of providers, assisted living and active adult, will offer both types of products and services as the lines further blur. Both terms "assisted living" and "active adult" probably are destined for the scrap heap as the architecture and services change to accommodate boomers. Whatever new terms come into favor probably will not involve any reference to old age, frailty, illness, or limitation.

Trend #9: People will Create New Types of Communities Themselves if They Can't Find What They Want

Ruth Glendinning, who calls herself a cultural strategist, couldn't find what she was looking for, a nice assisted living facility for her father. She looked at many of them. "Every one we found was either a linoleum ghetto or a velvet cage," she said. So she and her husband Keith are creating their own new type of community centered around wellness in combination with assistance services. They are calling it WALT, Wellness

Assisted Living of Texas. (Walt was also the name of Glendinning's father.) They are pioneers in the area of what Schriener is calling Citizen-Created Communities (CCCs).

What Glendinning has in mind is to build three duplexes with a caretaker cottage. Assisted living services will be there for the residents and the community shares some of the tenants with cohousing in that residents will share in the care of gardens, service pets, and each other. Solar panels and other green construction will prevail in the green-friendly Austin, Texas, location.

WALT will partner with a local emergency medical service (EMS) and a local nursing school so residents will get care at a rate much reduced from what 24/7 nursing care would normally run. The care will be there if needed but the focus will be on wellness with nutrition, water therapy, and proximity to shopping that's easily walkable. "We're creating a place where we want our dad to be, and us!" she said.

Schriener checked in with Glendinning in December 2009 and she said, "The project is in concept development stage and the business plan is well under way. However, the land has been purchased, designs are being discussed, and we have already had serious interest from future tenants, investors, and even the state of Texas." She added, "Nobody has any idea how big the wave is that's about to hit the shore. It's just now hitting the edge of people's psyches that even if they can't change *where* they live, they can change *how*."

If Glendinning can create her own new type of community, so can others. Boomers are inventive, innovative, creative, and driven. So there will be more CCCs, you just wait and see.

Trend #10: One of the Biggest Trends in the Next 5 Years May Be Something We Don't Yet Know Anything About
Going back to the concept above that Taleb wrote about in *The Black Swan,* something that has the biggest impact on the way we live probably will happen in the next 5 years, and we are clueless as to what that might be. It won't necessarily be bad, either. Look at how social networking—Facebook and Twitter, for example—went from nearly zero to 60 in about a year. Just a relative handful of people knew anything about social networking one moment and the next it was totally changing the way people relate to each other. Go figure. Something like that, or like 9/11, if we're not lucky, will happen again, and our lives will forever change. We just don't know what that "something" is.

Also Considered
The authors considered naming the remodeling business which was outpacing new home construction but that too has finally fallen victim to the recession. Early in 2010, fledgling signs of recovery were starting to emerge, but, as with new home construction, it will take awhile before this segment fully recovers. The authors also looked at modular or prefab construction but that business has drawn back to its base of manufacturing HUD homes again. Michelle Kaufmann Designs, perhaps the highest quality producer of modular homes, has closed its factory and is out of that business for now, focusing on design. Kaufmann's, *PreFab Green*, was published in 2009. Storm-resistant design has made trend

lists in coastal areas but not elsewhere. Closets that are a little larger are not really a trend. The authors don't see the data to support the idea that universal design is becoming a full-blown trend as of now. The fact that it makes sense is not enough, because there is less economic benefit than there is with energy conservation. Someone mentioned flexibility in floor planning, which continues but doesn't really rise to the level of a trend. Also rejected: Design accessories that reflect nature, car lifts in garages, showers instead of tubs, dark wood floors and woodwork, mixing kitchen cabinet colors, saturated colors, cheese coolers for the Formaggio appassionato, his and hers amenities, and mirrors to add glamour. Pockets of builders will implement the foregoing and individuals will request them as well as hundreds of features that inject personal expression into their homes, but that doesn't make them trends. Maybe the idea that so many different special requests are forthcoming is a mini-trend in itself. Most of the rejected ideas were not exceptional enough to be meaningful to builders looking for ideas to use in their new homes.

CRYSTAL-BALL GAZING

What would an aging-in-community model look like? Would it take a new urbanism form? Or would it be a modified lifestyle community of recreation and activity? Or would it be more like cohousing? Or would it be something entirely new?

As both the new urbanists and cohousing advocates believe, places for people to gather and connect are critical. Lifestyle communities also provide many such places, but the structural form of new urbanism and cohousing communities increases the number and size of these public gathering places. Both community concepts reduce distances between homes across streets both so people can interact and attempt to lessen the impact of the car on the living environment. Not everyone wants to be so close to their neighbors, preferring instead more distance and privacy.

Remember that baby boomers are hardly one large buyer segment acting in unison. Boomers are incredibly diverse. Look at their widespread ages, not to mention the differences in their income, background, cultures, and lifestyles. As Boomertising CEO Priscilla Wallace emphasizes over and over, "It's all about lifestyle. It's ALL ABOUT LIFESTYLE!"

Many boomers like where they live just fine, while some are ready for new experiences. Many don't know what they want until it stares them in the face. Nobody knows for sure, since this is unchartered territory, but the authors think that more boomers will look for a new way to live than their parents' generation did. Active adult communities as they currently exist are in jeopardy, partially due to the all-too-common isolation away from the mainstream of life. Some people now over 50 have young children, whether their own or grandchildren that they are raising. Many who do not have children at home like the normalcy and diversity of having kids in the neighborhood. Del Webb's 2003 "Baby Boomer Report" indicated that 75 percent of the younger boomers preferred a community that has a mix of all ages, and subsequent reports agree. This last point has already been recognized in intergenerational communities with what is typically an active adult community design that are springing up across the country.

Speculation Based on Forces of Change

Reporting on where the business of building for boomers is going from here is short on facts and long on conjecture, speculation, and pure guesswork. As in any down market, some things are working in some parts of the country. As far as the general direction the recovery will take, we can only rely on our observation of the forces of change:

Government: In this circumstance, unlike previous ones, government is and will continue to play a big role. But those in the housing business still harbor a natural suspicion of anything involving the government. Nevertheless, with government now in the financial sector as never before, more regulation and scrutiny are sure to follow.

Immigration: This has always been and will continue to be a major force for change. In some ways, such as the cost of housing, the United States is in a good place to be attractive to those outside the country. But jobs or the lack of them can be a drag on the growth of immigration. Cultural influences shape the lifestyles.

Aging of the population: This whole book is about that.

Religion/spirituality: In India designers are rooted in spirituality. Many homes have to have the kitchen on the southeast corner, the temple, bedrooms, and the openings into the home all in a certain place. Therefore, there's no way that they can do what we do in the United States when we build in quantity, that is, use the same floor plan if another home faces a different direction or map out a floor plan and then flip it to use in a home on the other side of the street or building.

Other forces to factor in when looking to the future:

- The growing income gap between the haves and have-nots.
- The fading into the past of the traditional nuclear family.
- Continued focus on conserving energy and all of the related consequences. Issues concerning the environment may be more of a social issue than a physical one.
- *Globalization*—anything can happen anywhere in the world and we feel it in the United States.

Schriener has observed for two decades while covering the construction industry that some things that appear incredibly obvious in terms of benefits actually encounter such resistance from designers and builders that years go by before they're even receptive to any kind of change, let alone do they implement it. It could be that they don't see the forces of change coming at them, or they don't think the change is worth the investment, or they are just too comfortable doing things the way they've always done them. As Schriener's late father said all throughout his life, "The only thing constant in life is change." We just don't always know exactly what will change or how fast.

What Kephart envisioned when he also first thought about what to put into this chapter was this carefree skipping around the world, our playground, if you will, but that has changed. Without falling into cynicism again, it's still possible to do anything we wish, go anywhere we desire, etc. All that has changed may be our mode of travel and the way we interact with people. This may be all for the better, but at worst it is at least as much fun as before and likely a lot more. His point is, the world has not changed, only our perception of ourselves within it. If we are sad we see sadness, if we are having fun we remember those places with great fondness.

The Long View

You look beyond whether great rooms are going to be more popular than living rooms and look at the major forces behind the scenes. Futurist Andrew Zolli said we make the mistake of reacting to the obvious things that happen quickly instead of the more slow-moving things that are more hidden. The recession didn't hit overnight. There were people who foresaw it. Terrorism didn't crop up overnight either. Factors were at work in both cases that moved slowly and were not immediately obvious. The problem is that we're all looking for what will work for us in the next 6 months or a year. That is the wrong formula for the near future. Buy a piece of dirt, build something, and rent it. The idea of selling something sure sounds as if the rental market is going to far exceed the for-sale market for an unknown number of years, at least 3, which is way beyond where most of us are looking.

Boomertising's Wallace said she sees boomers headed into some potentially dark times, or at least challenging. She said: "For boomers, nothing is secure. They can't rely on their kids, the government, or their own pocketbook to support them like their parents did and do. That's really big. The big deal about it is that with all of those things, the rug has been pulled out from underneath them. Time's running out and there may almost be the necessity of what looks a lot like communal living. If they can't hire from the outside, maybe they can exchange with each other."

REFERENCES

1. Administration on Aging Web site. www.aoa.gov. Accessed August, 2009.
2. Baker, Kermit, "As Housing Market Weakens, Homes Are Getting Smaller," *AIArchitect*, Washington, D.C., June 6, 2008, available at http://info.aia.org/aiarchitect/thisweek08/0606/0606b_htdsq2.cfm. Accessed December, 2009.
3. Cobb, Rodney L. and Scott Dvorak, "Accessory Dwelling Units: Model State Act and Local Ordinance," AARP Public Policy Institute and American Planning Association, Washington, D.C., April 1, 2000, available at http://www.aarp.org/research/ppi/liv-com/housing/articles/accessory_dwelling_units__model_state_act_and_local_ordinance.html. Accessed December, 2009.
4. Del Webb, "Baby Boomer Report," Pulte Homes Inc., Bloomfield Hills, Mich., 2003, 2010.
5. Jayson, Sharon, "Rise in Multi-generational Households Presents a Changing Family Picture," *USA Today*, June 26, 2007, available at http://www.usatoday.com/money/perfi/eldercare/2007-06-26-elder-care-generations_N.htm. Accessed December, 2009.
6. "Paying for Long-Term Care," National Center for Assisted Living Web site, http://www.longtermcareliving.com/financial/pay/. Accessed December, 2009.
7. Stewart, Martha, "My Mother's Legacy," The Huffington Post, December 1, 2008, http://www.huffingtonpost.com/martha-stewart/my-mothers-legacy_b_147527.html. Accessed December, 2009.
8. "The Cost of Assisted Living," Assisted Senior Living Web site, http://www.assistedseniorliving.net/ba/facility-costs.cfm/. Accessed January, 2010.
9. Rogers, Tracy, "Granny Flats," *Arkansas Democrat Gazette*, September 12, 2009.
10. Sage Computing Inc., "Accessory Dwelling Units: Case Study," U.S. Dept. of Housing and Urban Development Office of Policy Development and Research, Reston, Va., June 2008, http://www.huduser.org/portal/publications/PDF/adu.pdf. Accessed December, 2009.

11. Taleb, Nassim Nicholas, *The Black Swan: The Impact of the Highly Improbable*, Random House, New York, NY, 2007.
12. U.S. Census Bureau data, 2000–2009.
13. Wylde, Margaret, *Boomers on the Horizon: Housing Preferences of the 55+ Market*, BuilderBooks (National Association of Home Builders), Washington, D.C., 2002.
14. Wylde, Margaret A., "Active Adult Housing Prospects," National Association of Home Builders Webinar, September, 2009.
15. Wylde, Margaret A., *Right House, Right Place, Right Time: Community and Lifestyle Preferences of the 45+ Housing Market*, BuilderBooks (National Association of Home Builders), Washington, D.C., 2008.
16. Zolli, Andrew, Keynote Speech, Build Business Conference, Society for Marketing Professional Services, Las Vegas, Nev., July 18, 2009.

11

Determining and Developing Your Niche

"One thought driven home is better than three left on base."

—James Liter

Today, more boomers will look for new ways to live than did their parents' generation. Clearly no single home offering will appeal to all 77 million boomers, given their diverse wants, needs, wealth, and cultural history. This gives you the opportunity to create a niche, a certain type of home or neighborhood to specialize in rather than trying to build something that's satisfactory to a great number of people but not thrilling to very many. If you are new to designing and building for boomers or even seniors, you may not know where to start in determining what niche would best suit you and your market. If you get it wrong, it's an expensive mistake, even fatal in the business sense. You may be an excellent designer or homebuilder and you may have been very successful for many years, but if building for boomers and aging adults is new to you, you really need to do a lot more upfront thinking and planning than you are used to. The more information you have regarding your potential market in your particular location, the better off you will be. And, if you think you can do pretty much what you've been doing, just for an older market, learning otherwise can be costly.

One very successful custom homebuilder in central Maryland decided to develop a small age-restricted project for the active adult market. His previous market in which he had built his reputation for over 20 years was intergenerational. He procured a parcel of land not far from shopping and services, built a small building that served as a clubhouse, built and merchandised five model homes, and began building more homes on the parcel. The clubhouse contained, basically, a large open space that was to be used for anything and everything. It was only accessible with a keypad and code. If a resident in the complex had a private event, there was no way to let anyone inside the clubhouse know when guests had arrived. The homes were beautiful but overbuilt, much like the custom homes the builder was used to constructing. Homes were very close together and the interior streets were so narrow they would not accommodate two cars going opposite directions. There were no pools or green areas planned for the land inside the horseshoe; homes were to fill the entire space. Only one of the five models had a standard elevator; each of the others had an area blocked out for an elevator, which could be added for an additional $28,000 or so. Not surprisingly, the homes were not selling all that well before the real estate market tanked, and the recession did in the builder, who got hit with lawsuits, had to go to auction with 30 homes in the development in an effort to pay debts, and ended up abandoning the project, closing his office, and filing for Chapter 11 bankruptcy protection. Driving by and through the development late in 2009 was a depressing venture. A few homes were occupied; some completed homes were empty; some unfinished homes stood like half-dressed mannequins in a store window; two lots had open, incomplete foundations dug, which lay open, presenting a safety hazard and filling with rainwater and who knows what else; and half of the land was untouched. It's not a guarantee that this builder would have succeeded if he had done more homework to learn what to build in an age-restricted community to sell more homes, but his huge miss there certainly had to have contributed to his downfall.

That story is repeated in many cities throughout the country. When times are good, builders can get away with being off the mark quite a bit more than when times are tough and people are only going to move if they get exactly what they want. One thing the recession has done, which several builders large and small have told the authors, is to give builders the incentive to reassess their businesses, their approaches to the market, and their offerings. After initially reacting negatively to the economic plunge, many of them are using the opportunity to step back and retool their businesses, even if they downsize, and are emerging stronger than ever in terms of focus and having the right homes for the right market.

So how do you determine your niche and fulfill the requirements of that niche?

QUESTIONS TO ASK, STEPS TO TAKE

Asking Questions

Answer the following seven fundamental questions and you will have a pretty good idea of who your market will be and how big that market is. The point of the exercise is not to prejudge the outcome. The answer will reveal itself to you in the end. Just because a builder decides to address a certain niche does not mean it makes sense with his location or in his market area. The answer may not be what he wanted but will more likely be what will work for the market that is there. Preconceptions will only lead to misdirection and failure unless luck steps in and saves the day. The seven fundamental questions are

1. Where and how large is your market in terms of total households of all ages and sizes?
2. How many households in your market area are in the 50+ age group?
3. How many of those 50+ households are likely to move in the near future?
4. Who is your buyer? You need a clear vision of the families you are designing your community for.
5. How much can your target families afford to pay for a new home in your community?

6. Is your location going to appeal to all of your targeted buyers or only a portion of them?

7. What kind of a community will your market prefer? Are they interested in a gated environment or one that is intimately connected to the larger neighborhood?

Research can help you answer those questions about your local market. After you have the answers to those questions, then you can follow up with more detailed investigation into lifestyle choices, age restriction, the role of religion in their lives, and their personal preferences in their homes. This process of refinement will continue through the entire time you are marketing and selling the homes within this community.

Questions that should be answered at almost every step of the way are whether you should continue on the path you've chosen, change direction a little or a lot, or pause and reconsider. Markets change, the economy changes, and your target market may shift as well. Sometimes this is good. Artisan Homes had that issue with its Artisan Lofts in Downtown Phoenix, which originally were targeted to young couples but ended up being very appealing to empty nesters, and the company responded with a more inclusive marketing message.

Taking Action

The first thing you should do is determine how big a market you're working with. It won't be as big as you think. The world is not your market, or the whole United States, or your entire state, or even your town. You need to take some steps to determine your likely universe of prospects.

Step 1: Define Your Market Area
The days of drawing a 5-mile circle around your community location are past. Transportation corridors and local population movement patterns play a strong role in defining your market area. Researcher Susan Brecht, in her article "The New Active Adult Community: Defining Feasibility," said, "Defining the market area has become an art ... and that market area is the structure for the rest of your study." You need help when you go to establish your market area and determine the locations where your future homebuyers currently live. This is not something you should do alone. You should do market research, either formally with a good researcher experienced in the 50+ market or with the best local marketing expert you can find.

Step 2: Compute Your Market Size
Let's assume your market area as you have outlined it in Step 1 encompasses a total population of 100,000 people. About 25 percent of them likely will be over 50 years old by national averages. This equates to 25,000 individuals, but you should verify this with your local market researchers.

Step 3: Estimate Those Likely to Move
Del Webb, still the leader in active adult community developments, acknowledged in its 2003 Baby Boomer Report that only 7 percent of the 77 million baby boomers are likely to purchase one of its homes. According to that same report, 59 percent of those boomers are likely to buy a new home sometime in retirement. Seven percent of 77 million

equates to 5,390,000 people. That means Del Webb will have 3,180,100 families with at least one 50+ plus member in it who will be interested in the lifestyle Del Webb's communities offer. That equates to 159,000 couples or families per year over the next 20 years. Considering that Del Webb had built and sold only 80,000 homes before 2000, this is a sizable niche it has to serve, even though competition is growing.

In fact, Del Webb is expanding into the intergenerational and non–age-restricted types of projects to serve more niches, and, according to Robert Powell, who writes for *MarketWatch*, by 2009 they had some 50 or so communities in the snowbelt cities. Many of its newer communities are smaller than its typical Sun City development of 4,000 to 6,000 homes. Some, like Anthem Parkside in Phoenix and Anthem Ranch in Denver, are intergenerational with limited or no age-qualified areas within those communities. All in all, Del Webb's chosen niches look very good for the company to ensure future growth as the baby boomers come of age.

You are not Del Webb or Pulte, so let's figure your probable market:

1. All 77 million baby boomers will be over 50 by 2015.
2. Around 50 percent of boomers are likely to move to a new home in retirement, some 38.5 million age 50+ homebuyers (minus about 10 percent, says the 2010 report).
3. Between 20 and 25 percent of the U.S. population of 300 million people are over 50 today, and that percentage will increase every year.

Therefore, using the 25,000 people estimated above in Step 1 and 50 percent as in the Del Webb example, in your market 12,500 people over 50 are likely to move in retirement. This works out mathematically to approximately 7,250 families.

Step 4: Define Your Targeted Family Unit
A little over half, 53 percent, of 50+ households are composed of two people, usually married couples; 34 percent of households contain only one person; 7 percent contain three people and 6 percent contain four or more.

If we choose the two-person household as our targeted family unit, 53 percent of 7,250 families equals 3,845 families.

Notice that as you begin to define your niche the numbers grow smaller but the likelihood grows stronger that those targeted buyers will choose your home design (see Fig. 11-1).

In reality, by offering several model home designs a builder can design each to appeal to other buyers such as singles and households of three or more. However, you will probably want to offer several models to your targeted family unit to increase the chances that they will buy, and if you use up some of those models for other family units, you will have fewer to offer the group you really want to sell to, so consider the effects of the trade-offs before making your decisions.

Step 5: Target Household Income
About 65 percent of homebuyers age 55+ report a total annual income of more than $50,000, according to a study by researcher Margaret Wylde. Choosing the two-person

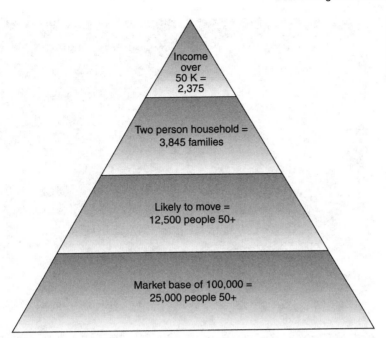

Figure 11-1
Niche definition
chart. (*Drawing by
Mike Kephart.*)

Income
over
50 K =
2,375

Two person household =
3,845 families

Likely to move =
12,500 people 50+

Market base of 100,000 =
25,000 people 50+

household as our target market reduces the number of potential buyers but narrows down the niche to 2,500 ever more eager families. Obviously, in large communities you can include homes for people in other income groups but this is more difficult in smaller communities.

Step 6: Factor in Community Location

Thirty-two percent of the 55+ homebuyers in Wylde's study said they prefer a close-in suburban area; 30 percent prefer a rural location; 33 percent prefer the outlying suburban area; and only 5 percent prefer the central city.

Whichever location you choose, your community will lose appeal to those who prefer the other locations, but take heart! Those who remain will find your community especially appealing.

Location is a niche driver that cannot be broadened easily. Some buyers can be drawn to a development by strong design even if the community is not in their preferred location, but it's making your job much more difficult.

One project drew buyers from large homes in gated golf course communities to relocate on tiny lots in a development surrounded by office buildings and having few amenities. Some buyers wanted to move down and others liked the location but the strong design appeal of this high-quality community was attractive to others that would not have chosen this community for any other reason, including location (see Fig. 11-2).

Providing more of the other qualities home shoppers seek in a community can help stretch your market area but putting too much faith in this effort can be risky.

Figure 11-2

Downtown high-rises will not appeal to everyone but will be very exciting to a few. (*Photo courtesy of KEPHART.*)

Location is a powerful element of community selection by homebuyers. Let's play it safe and choose a close-in suburban area. We may attract some buyers from the center of cities who can't afford the price tag for a loft, and we can certainly draw from those wanting outlying suburban areas, as long as we can keep prices competitive with outlying locations.

Completing this step, we calculate that 32 percent of our 2,500 highly qualified families equates to 800 families remaining in our niche. *These families represent buyers who want our homes, like our location and can afford what we offer.*

You need to consider competition as you move through the exercise. Eight hundred families as a highly qualified market is a huge number if you are only building 50 homes and have no competition. You can continue fine-tuning your market niche. Decisions about home type, single-family or attached, or whether to gate the community will all narrow your market niche even more. Keep this in mind: *The tighter you define your market, the smaller the niche becomes and the more excited the buyers will be with your offering.*

Our hypothetical niche is right at 800 households. If we want a larger market we could go back and analyze the potential for a broader, non–age-restricted buyer group and redefine our market area and the purchasing power of the buyers we chose to serve.

This oversimplified exercise provides you with an idea of how these design decisions both strengthen the market appeal of your community and reduce the size of your potential buyer group, thus making it easier to target your marketing efforts to those within your niche. You can narrow down your niche as far as you want (see Fig. 11-4) as long as you realize that the narrower your niche is, the fewer potential buyers there are. Even if those you target are thrilled to find exactly what they want, enthusiasm doesn't make up for lack of *enough* potential buyers in your target market.

There was actually a community started in Florida for retired professional sports figures. Billed as "the world's most exclusive retirement community," ex-Big League pitcher Sam McDowell had a dream of a community populated by his peers, "The City of Legends." "Sudden Sam," as he was known when pitching for the Cleveland Indians, retired in 1998 and began work on his dream in a location west of Orlando. Originally conceived as a haven for professional athletes who had suffered psychological, physical, or addiction problems, today "The City of Legends" is an umbrella concept uniting new and existing developments around Clermont, Florida.

Niches for the 50+ boomers may be found everywhere: small towns and rural locations; family compounds; second home communities; spiritual groups; eco-conscious towns and communities; social groups; openly gay communities; communities linked to colleges and in more traditional locations such as downtowns, around transportation hubs, in suburban shopping centers, and in old residential neighborhoods.

Caution! The Federal Fair Housing Law is clear. No residential community for sale or rent may discriminate against anyone due to their race, religion, gender, age, etc., with the sole exception of the allowable exemption for those over 55 years of age.

EXAMPLES OF POTENTIAL NICHES

Location-Based Market Niches

These are 50+ development ideas that are location-specific. Some are naturally occurring, such as the return to *downtowns,* and, to the other extreme, to *small rural towns.* Many 50+ boomers and others are strongly attracted to the promise of a quieter life at a slower pace than they can find in the city.

Downtowns are attracting a significant number of 50+ homebuyers. They are moving back downtown, choosing contemporary lofts and more conventional condos, but the common draw is the action, the culture, sports, entertainment, restaurants, and the ability to walk everywhere they need to go in a single day. Between 1990 and 2000 the population of 64- to 75-year-old persons in downtown Chicago rose 17 percent. Austin, Texas; New Orleans; and Los Angeles have seen double-digit increases as well. A Del Webb survey of baby boomers indicated that 30 percent of those considering purchasing a home in an age-qualified active adult community prefer a community in an urban location.

Small towns have lost much of their economic base and have seen their populations shrink as kids mature and are forced to move away in search of work. As boomers relocate to these communities in search of a more relaxed lifestyle, new businesses, restaurants, and stores follow. Madison, Georgia, "the town Sherman refused to burn" on his march to the sea during the Civil War, just 50 miles east of downtown Atlanta, has all of the necessary ingredients. The town's history is rich and colorful, its restaurants are many and varied to suit all tastes, and its economy is strong and healthy.

Transit-oriented designs (TODs), are created by urban and city planners. TODs are covered in more detail in Chap. 3. The transit system created the opportunity for infill and redevelopment, and thus TODs spring up spontaneously around mass transit stations.

Many 50+ boomers are attracted to *colleges* and *universities*, whether to resume lifelong learning, help teach or assist at the school, or be close to the youth, vitality, and intellectual stimulation of a college atmosphere. Most schools encourage the trend, seeing a potential source of income and talent.

The *old neighborhood* is about people staying connected to the neighborhood they know and love. Author Kephart could relocate to a TOD in Denver and only move 10 or 15 blocks from the single-family house he has lived in for 30 years. He could also move downtown to the loft he dreams about and still be less than 5 minutes from his old neighborhood and even closer to the restaurants and stores he now frequents. A move to a university town would mean leaving the city he knows and loves for an unfamiliar place. A move to a rural town would be similar, though many are less than an hour from where he lives now.

These are some of the considerations your buyers are faced with. Those who want to stay in familiar surroundings would be less likely to move to a university town or rural town where they would have to meet new friends but the more adventurous may try all these choices at some point. The location-based niche is something you give for a few rather than trying to satisfy everyone.

When you choose a location you have begun to define your niche.

Social and Cultural Niches

Four divergent community design concepts fit within this category:

1. Spiritual or religious
2. Cohousing
3. Lifestyle of health and sustainability (LOHAS)
4. Gay and lesbian

It is acceptable and legal to encourage buyers of a certain sexual orientation or religious belief to come and buy or rent homes in your community as long as you make it abundantly clear that everyone is welcome to live there. Religious-affiliated nursing homes, assisted living and continuing care retirement centers have a long history in our country, so don't let fear of discrimination turn you away from an idea that is close to your heart. Make yourself familiar with and follow the guidelines of the Federal Fair Housing Law and go from there.

There is little difference in home designs for these cultural niches. A person or couple's sexual preference, religion, race, or commitment to social cause has little impact on how they live in their homes. A few exceptions to this broad generalization do exist, though. Homes in cohousing communities are designed to connect strongly with the pedestrian street to facilitate social interaction from inside the homes to the people outside, and also some religions encourage believers to incorporate prayer or meditation spaces into their homes.

For example, many of Del Webb's newer communities are smaller than its earlier ones. Corte Bella, a gated country club adjacent to Sun City West in Arizona, has only 165 homes and is priced from about $250,000 to $590,000. It is Del Webb's first active adult country club community. Del Webb bills its Solera communities in several locations around the United States as "a collection of neighborhoods." In some of its new communities, Pulte blends Del Webb with the Pulte brand and, in the first such venture, DiVosta Homes, the third Pulte building company of "the Pulte Family of Companies," is included at Ave Maria near Naples, Florida. Ave Maria will have 7,400 homes in four distinct communities. The communities are planned around Ave Maria University campus, to integrate lifelong learning into the residential communities. An oratory for Catholic Ave Maria University was designed to be a signature feature and central focal point for the town center. Tom Monaghan, founder of Domino's Pizza, has created a powerful vision for this new town. Here is part of the text of the vision: "Ave Maria, Florida, is a new concept—a place where living and learning form an integrated whole, where sustainability and community are complementary values. Here, families, students, and retirees of any race, religion, or ethnicity will live and thrive together in a beautiful and safe environment. Neighbors will care about neighbors, friendships will span generations, and all will share a sense of pride. Ave Maria will be a real hometown." Ave Maria is an example of two cultural niches plus two lifestyle choices: the university link, the spiritual overtones, the resort community atmosphere, and the intergenerational mix.

Spiritual

The new "boomer American dream" includes top priorities of being true to self, not selling out, and achieving inner satisfaction, versus power, influence, and wealth. Most current examples of spiritual communities, such as ElderSpirit in Abingdon, Virginia, and Silver Sage in Boulder, Colorado, are small cohousing developments or intentional communities.

Ave Marie, as described above, is at the other end of the spectrum. Despite combining several elements, the 5,000-acre development centers around a new Catholic university and a 1,300-seat oratory (cathedral) (see Fig. 11-3). The completed town will undoubtedly create the perception of a religious community by virtue of the Catholic-affiliated university and the dominating presence of the huge, 100-foot-tall oratory in the center of La Piazza, the town square.

Cohousing

As you saw in Chap. 3 on neighborhoods, a cohousing community is a cross between an enclave, a village, and a commune. Homes each have their own bedrooms, living rooms, kitchens, etc., but are clustered around common or shared facilities and amenities, often including a mega home of sorts that includes a super kitchen so residents may take turns cooking for all and a dedicated playground just for children of residents of the

Figure 11-3
Oratory at Ave
Maria in Naples
serves as the
central focus for
the community.
(*Drawing courtesy
of Barron Collier,
Brian Gable.*)

cohousing community, some of whom may share childcare responsibilities. Cohousing is a relatively new way to tend to the time-tried tradition of people striving to house themselves with independence, dignity, safety, mutual concern, and fun.

Lifestyles of Health and Sustainability (LOHAS)
Public awareness of the real need to live a more sustainable life grows each day. Most people believe they have a duty to protect the environment for their children and grandchildren. Energy savings, part and parcel of living in a sustainable fashion, is also a money saver, making it the top priority for many 50+ homebuyers. Living simply in a smaller home is an easy sustainable technique, as is living closer to workplaces, shopping, entertainment, and recreation. There are more than 20 eco-villages around the country with more being formed each year.

Lifestyle Market Niches

This group of market niches is all about *how* people live in a particular community. They can be in any location that makes sense and they may exist in combination with one or more social or cultural niches.

Age-restricted active adult communities are those where only people over 55 years of age are accepted. The large ones and even some of the smaller ones have something for everyone: a whole range of activities and events planned by a staff, including a social director; health and fitness facilities; sports; golf—you name it, it's there. This is a familiar community model that still works but for less than 10 percent of the 50+ and maybe for even fewer boomers.

Exclusivity is also an old model that retains a loyal following of the privileged who want that private lifestyle.

Second homes, of course, aren't new but boomers have a different view of them compared to members of their parents' generation.

TNDs, or *traditional neighborhood developments,* are relatively fresh on the scene but until recently did not provide 50+ buyers the types of homes they want. The community concept is perfect with a mix of uses and housing types and a network of public spaces leading to a village center. Chain restaurants tend to prevail but some locally owned eateries also are usually within walking distance. The maintenance-free features of the community enhance their appeal, but two or three floors of living space and flights of stairs up to front porches have turned away many 50+ buyers.

Walking-oriented communities are well suited for boomers. Many 50+ buyers enjoy walking and for some it needs to be the only community amenity. Imagine a complete community designed around walking. No, the new urbanists didn't invent walking, but they did remind us that walking is an important quality in the design of a community. There could be forest and nature trails, walkways that allow socializing, walks to retail and business areas, long loops, short loops, and trail systems expressly for exercise, similar to the trails in Rock Creek Park in Washington, D.C.

Intergenerational communities may be the dominant models for community design in the near future. A high percentage of 50+ homebuyers like the diversity and the presence of families with children, and builders gravitate to it naturally. It's not the same old sub-division model, however. Fifty-plus buyers have put their stamp on a truly innovative community model that is good for buyers and builders alike. Homes are no longer designed exclusively for families with children. Single-story homes are offered as part of the product mix, and lots are smaller. Communities are more pedestrian-friendly, which means more compact for shorter distances to everything and with walkways that connect residents to the world outside their neighborhood. Recreation amenities and commercial services are integrated more finely with homes and apartments.

Cohousing presents opportunities to create niches within the cohousing niche. Some kind of common interest or characteristics draws residents together; they want to be very close to others of the same ilk. Charles Durrett mentions in his book *Senior Cohousing: A Community Approach to Independent Living* that cohousing developments exist that are based on spirituality, aging in community, the environment, sexual preference, and others. These niches-within-a-niche drill down deeply into the market looking for that special lifestyle. Cohousing communities typically cluster 20 to 30 private residences around common spaces and facilities. The common house is the heart of the community and it serves as a gathering place for group activities, meal preparation, common dining, and meetings. Due to the small size of cohousing communities, this form of development is very well suited for custom building companies and other smaller-volume builders.

Gay and lesbian communities draw people who prefer to live with others with the same sexual preference. For people in the market for a 50+ gay and lesbian community right now, choices are limited but more selection should start to appear in the near future. Gays and lesbians will make up about 7 percent of the 50+ housing market in a few years, a significant enough number to attract the attention of developers and builders. Joy Silver in an article on www.gay.com said, "Standard retirement can mean isolation for lesbians and gays, who may find themselves leaving West Hollywood or San Francisco for the predominantly straight retirement or assisted living facilities that have been the only options until now." One community, Rainbow Vision in Santa Fe, New Mexico, opened in 2006 with an offering of 40 "club casitas" and 20 condominiums, all fully owned by the home-buyer. The complex also includes 60 independent living units and 26 assisted living units for rental fees that include utilities, services as selected, amenities, dining, and healthcare. Rainbow Vision is close to two Santa Fe colleges as well as shopping and restaurants.

Going back to school is a long-term desire of many boomers. Living close to a university is ideal for an active, stimulating lifestyle and being around young people. Universities offer special involvement in classrooms through teaching or volunteering. The libraries and research facilities are top notch and entertainment is plentiful from plays and concerts to lectures and sporting events of all kinds. One of the features of life on a campus that is a major draw for active adults is the culture of walking. Coffee shops and restaurants cluster nearby, as do all of the necessary services from dry cleaning to shoe repair. Campus plans are uninterrupted by streets or roads, making walking convenient, safe, and enjoyable. Over 60 university-linked retirement communities exist currently and several more are under construction or on the drawing board.

Accessory dwelling units (ADUs) are a new solution for a challenge for baby boomers that will be increasing from now on. Many boomers still have children at home and they also have aging parents who are beginning to need care and a safer housing environment. Housing a parent in an independent living community is expensive and costs for assisted living are even higher. These boomer families are in a crunch between supporting their children and their parents, earning them the nickname "the sandwich generation."

ADUs are small backyard cottages sometimes called granny flats, carriage houses, or, especially in the Southwest, casitas. They have been around for decades, having originated in times when household servants were common. They are becoming more popular now as they present a solution for boomer families to take care of aging parents as America's aging population continues to swell or for adult kids who come back to live at home. Santa Cruz, California, has been a leader in housing its elderly in ADUs added to existing single-family lots. As noted earlier, in the interest of full disclosure, author Kephart's Sidekick Homes is a relatively new firm specializing in ADUs. He is spending a lot of time and effort working with municipalities around the country in an effort to get zoning rules changed to accommodate the little homes.

Each niche from the gay and lesbian with its 7 percent potential market (5.4 million people) to the downtown 17 percent (13 million people) can be equal big business for builders. Even small niches can make a successful business for you. For example, the size of the cohousing niche is tiny, but Jim Leach's Wonderland Hill Development Company in Boulder, Colorado, sells up to 200 homes per year in cohousing communities. And, with boomers on the horizon, each niche is sure to grow. See Chap. 3 for more detailed information on each of these niche neighborhoods.

Having a clear picture of your future buyers and how they live is the genesis of a vision. Once a vision is established, the team can flesh it out and illustrate the core ideas for all to refer to and use as their guiding principal for making design decisions and, as importantly, for marketing.

Figure 11-4
River runners could constitute a small niche. (*Photo Courtesy of Janice and Bobby Lipp.*)

DESIGNING FOR YOUR NICHE

Land Planning

The market is an important force in shaping community design. The land, its topography, vegetation, views, and location are other powerful forces. These forces must be working in harmony or you have either chosen the wrong market niche for your site or the wrong site for your market. This concept continues with each element of community design. For example, the architecture must also work with both the site and the market. One-story, easily accessible homes are difficult to design at compact densities on steeply sloping properties, and they are nearly impossible to design at a density of more than three or four homes per acre without serious trade-offs.

Guided by the vision developed earlier, the market study and the opportunities and constraints offered by the land, it's the professional land planner's task to create the bones of a structure upon which others can build a living, working community. That phraseology is to remind you to keep it clearly in mind that a development is not a "community" without the interaction and socialization of people going about the business of living their lives. You, as planners, architects, or builders, can only set the stage to facilitate the formation of a viable community.

The things you don't do may be the most important. Leaving special natural features alone, such as lakes, forests, natural grasslands, streams, or hills, can result in the most treasured open spaces in the development. They can even create the community identity for people to relate to and embrace as part of their home. Historic structures, such as classic barns or old mills, can do a similar job for community identity.

Tyson's Corner in Virginia, near Washington, D.C., is an example of a highly urbanized city named after a country store that was located at the intersection of two rural roads. The legacy of a rural life lives on as part of Tyson's Corner today. The White Fence Farm is a suburban community in Lakewood, Colorado, that kept its namesake white fences and popular family restaurant of the same name. Residents in that region of the city know the White Fence Farm. Whether or not the Woodlands near Dallas includes woods, it's safe to say that most people in Dallas know the Woodlands.

Identity can be developed from an association with a local landmark such as a bridge, park, square, mountain, or well-known neighborhood. That's more marketing than design, unless a physical reference to the landmark can be part of the design. Calling a community River Oaks would be a joke of a name if the development had no oaks and was nowhere near a river. Entry monuments, so often relied on to carry the lion's share of the load of community identity, is ill-suited to the task, particularly when there are multiple entries, as in interconnected neighborhoods. Simple, well-designed street markers and signs can do the directional work and can reinforce identity by their design.

After identity, it's the planner's job to begin building the physical elements of a community. Walking is so popular among the 50+ group—87 percent per Del Webb's report and it ranks number one in mature market researcher Margaret Wylde's studies—that it makes sense to start the planning by locating walks, streets, paths, trails, and pedestrian amenities before anything else. You and the planner should decide whether to locate trails adjacent to lakes and other features so the entire community can enjoy them or to put home sites in those locations so lot premiums may be increased. Giving everyone the opportunity to see and access treasured natural areas will increase the value of all homes in the community, but this is an important early decision for the design team.

After planning the systems of circulation for cars, pedestrians, cyclists, and maybe horses, comes locating places to gather. Here people can meet and get to know each other while shopping, attending an event, working on a volunteer effort together or sharing stories about golf games or fishing successes. Quality communities need many such spaces for the myriad of functions the residents may want to initiate.

You may have noticed that we are well into the planning and design process and we have not located a single home or community building. We're still not there because several more decisions must now be made:

- Will this be a gated community? If so, the number of access points will be minimized.
- Will this be an interconnected community? In this case there would be no gates and many access points to diffuse traffic and to connect the development seamlessly with the adjacent neighborhoods.
- Will price points and product types be separated or more seamlessly mixed? The typical master-planned community separates product types and price points, but TNDs and intergenerational communities are more seamlessly mixed.

You can see how important the vision can be when faced with these and hundreds of other decisions.

You may begin conceptualizing landscaping, storm water engineering, and utility design at this point. The design of these elements, while always important, is even more crucial in this era of green building and sustainable community design.

The final stage of land planning brings forth the most visible and memorable details. Street architecture, lighting, furniture, signage, and sometimes mailboxes are all the finishing touches to the "bones" of the community.

More Planning for Your Niche

You may initiate the entitlement process early in design, but most jurisdictions now insist that the entire community design be completed prior to the submittal of a final planning document. The entitlement process is often a lengthy and somewhat frustrating exercise. Depending on the jurisdiction and the political environment, the steps through the review and approval process may take from one to several years. This lead time, plus the months or years the design process takes, makes the reliability of prior research especially important.

Plans must include provisions for changes requested during the technical review process. Markets shift, split, merge, and evaporate, so you may need to employ new strategies. Common techniques used to provide for the eventuality of a market shift include documenting the ability to increase or reduce density; to change uses, that is, from commercial to residential, single-family to multifamily, etc.; to alter community amenities; and to change the phasing and time schedule of development. For the 50+ communities you might add being able to change from age-restricted to no restriction or vice versa; and to modify covenants and restrictions for later phases, allowing you to increase or decrease the level of common maintenance for later phases. For instance, an economic shift could render a community uncompetitive. It would be wise to plan for such eventualities. Home size minimums may be another possible modification in reaction to a market change, so be conservative in establishing minimums or avoid them altogether if possible. They hint at being exclusionary and elite, an attitude that elicits negative response unless done openly as part of the community identity.

Architecture for Your Niche

Author Kephart was three-quarters of the way through a 3-day design team meeting with a well-known active adult community developer and had not yet touched on the design of the homes, when the CEO came over and sat down next to him. The CEO put his arm around Kephart's shoulders and whispered what he no doubt thought were encouraging words: "Mike, I hope you understand—we are not selling homes. We are selling a lifestyle." He later elaborated, explaining that although the home designs are important, it was the clubhouse, the golf course and all the community amenities that they felt created the image they would be selling to the public. They were selling "an affordable resort" and about 13 percent of the country's retirees were moving to the Sunbelt to join in the fun. Much has changed since then.

Balancing Wants, Needs, Budgets

The typical post-war single-family homes were less than 1,000 square feet in size and had no garage. Today the median home size is larger than 2,200 square feet with two- or three-car garages and a basement or unfinished attic with options to finish. When asked in Wylde's *Boomers on the Horizon: Housing Preferences of the 55+ Market*, 75 percent of the home shoppers said they would prefer a smaller, higher-quality home, but something is preventing them from acting on that intention. The median home size kept increasing every year until 2009, when architects finally told the authors they were seeing it go down an average of 200 or 300 square feet per home. Either builders were not offering a real choice of smaller, better homes or buyers were untruthful in their answers to surveys. The truth is probably more complicated and personal than either of those choices. When it comes to design, the reasons for the choices people make are highly personal. A particular priority may cause a buyer to choose a larger home based on the configuration of rooms. One boomer couple had a budget in mind and narrowed their choices down to two homes. One was in a more desirable location, perfectly sized and right at their budgeted price. The other was several hundred square feet larger (which they didn't need or particularly want), was on a rather strange lot hidden from the street, and was priced significantly higher than what they had expected to pay. However, the larger one had two beautiful office spaces, and they couldn't pass up the opportunity for each to have their own office, so they chose that one pretty much solely for that reason. Now, 3 years after moving in, they are glad they did. Buyer choices are usually less rational and logical than we would hope. On the other hand, that can work to your advantage.

Good architectural design is a careful balance of visual, functional, and kinesthetic (feeling) elements resulting in the best possible home for a certain time, place, and family. The budget, a powerful constraint, is a family characteristic; cutting-edge technology, on the other hand, is a function of the home. Overemphasizing one aspect of design can negatively impact the whole. If the architect's concern with the appearance of the front facade takes priority, then window sizes, types, and locations may not work as well for the rooms inside. Conversely, if the design puts too much emphasis on floor planning and the location of windows inside the rooms, then the exterior design may appear unorganized and awkward. For example, when designing homes in the classic styles of colonial or Georgian architecture, windows may be scattered over the sides and the rear to allow the strict symmetry to happen on the front elevation and to provide necessary windows anywhere possible.

How people feel about a home or a certain space is concealed in their mind, often unknown even to them. For this reason, it helps to break down design into its elements to show how it works and how it imparts value to the finished home in so far as the architect can appeal to that rational part of the buyer's mind.

REFERENCES

1. American Association of Retirement Communities Web site, www.the-aarc.org. Accessed August, 2009.
2. Brecht, Susan, "The New Active Adult Community: Defining Feasibility," *50+ Housing* magazine, National Association of Home Builders, Summer 2006.

3. Del Webb, "Baby Boomer Report," Pulte Homes Inc., Bloomfield Hills, Mich., 2003, 2010.
4. Durrett, Charles, *Senior Cohousing: A Community Approach to Independent Living*, McCamant/ Durrett, Nevada City, Calif., 2005.
5. Powell, Robert, "More Older Americans Choose Age-Based Communities," *MarketWatch*, May 6, 2009.
6. Schreiner, Dave, "Small Niches Are Now Big Trends" presentation, International Builders' Show, Orlando, Fla., February 7, 2007.
7. Silver, Joy, "Rainbow Vision," www.gay.com, accessed 2006
8. U.S. Census Bureau data, 2000–2009.
9. Wylde, Margaret A., "Active Adult Housing Prospects," National Association of Home Builders Webinar, September 2009.
10. Wylde, Margaret, *Boomers on the Horizon: Housing Preferences of the 55+ Market*, BuilderBooks (National Association of Home Builders), Washington, D.C., 2002.
11. Wylde, Margaret A., *Right House, Right Place, Right Time: Community and Lifestyle Preferences of the 45+ Housing Market*, BuilderBooks (National Association of Home Builders), Washington, D.C., 2008.

Dos and Don'ts of Dealing with Boomers

"You're never too old to become younger."

—Mae West

CALL THEM THIS, DON'T CALL THEM THAT!

Baby boomers are used to being called that, or just "boomers." In Canada, they're sometimes called "boomies" and in Britain the boomer generation is known as "the bulge." But as a boomer turns 50 every 8.5 seconds these days, a question arises. What do we call them now? If you are dealing with boomers, the last thing you want to do is alienate them before you even get started on the transaction.

Call them "older," say they are in "midlife" or "middle age," call them "mature adults," "those over 50" or "people 65 and up," or keep calling them "boomers" or even "older boomers." Just don't call them "geezers," "curmudgeons" (except for those who strive to be regarded as such), "oldsters," "elderly," or "senior citizens," though "elder" is okay, according to a study of journalists who cover aging issues for the American Society on Aging. "Senior" is fine once they are over 70, but don't talk about their having a "senior moment" and use "old fart" with caution, even in jest, advised editors.

Boomers can't fathom that they got to be over 40, let alone over 50, let alone, for many of them, already over 60. They don't feel old. They still play tennis, golf, and softball. They go water skiing, snowboarding, hiking, and hang gliding. They run marathons, climb mountains, and go scuba diving. How can they possibly be old? So if you treat them as oldsters, they will be offended, and it's hard to sell a home to someone you've offended.

Boomers celebrated the 40th anniversary of Woodstock in 2009 like it was yesterday. They sat and stood and swayed to the music in concerts all over the United States put on by three 60-something-year-old original members of the Moody Blues in 2009, and they sang along, knowing every word, from the group's hits from 1968 like it was last year. They remember the past and they celebrate it from time to time, but they don't dwell on it.

For them the past is a place to visit like the grave of a dead cherished relative, not something to preserve like a wax figure in Madame Tussauds museum. They have moved on and they are so not the same people they were as young adults, so advertisers that try to appeal to boomers by playing groovy psychedelic music from yesteryear in their commercials risk putting off boomers who think that's how they are seen, as relics from a not altogether positive bygone era. Similarly, marketers of homes would do well to realize that boomers are looking ahead, not back, and communicate with them accordingly.

GENERATIONAL ROOTS

Fundamental to understanding how to deal with boomers is understanding the differences between today's boomers and the seniors we're all used to. Up until boomer babies grew into age 40+ adults, seniors were not a significantly large segment for marketers, who talked to them as a monolithic group, lumping them together by age and little else, noted Priscilla Wallace, president of Boomertising. However, boomers' huge numbers would put them into a completely different category even if there weren't additional factors to consider. Wallace explained that seniors—those silents and the Great Depression/World War II generation—by and large grew up relatively isolated; many didn't travel until later in life if at all. They grew up when luxuries and even some necessities were scarce, so even when they worked their way into abundance as adults, they held on to their fiscally conservative ways for fear of something happening to plunge them back into lack and sacrifice. The melting pot of America didn't have the numbers of immigrants it has now so the population wasn't as diverse as it is now. "Through the Depression the family came together," said Wallace. Seniors who were young adults during World War II tend to think of the past as "the good old days" and seek to preserve it. They yearn to go back to a simpler time when values were more clearly segregated into right and wrong. They don't understand or sanction letting technology do everything for them, though they appreciate it once their boomer children or Gen X or Millennial grandchildren teach them how to use it. For example, many of that generation either don't have cell phones (and don't see why they would need them) or only keep them for emergencies. For boomers, cell phones are a downright necessity, and not just for emergency calls.

Boomers, in contrast to seniors before them, grew up in abundance and are used to getting what they want and plenty of it. They grew up with the best of everything and tend to equate what their parents would consider luxury with necessity, which Boomertising's Mike Lynn calls "vanicessity." They traveled as teenagers and young adults and they could afford to; they experienced the world as a more open, accessible place than did their parents and grandparents and they became more globally aware and less fearful. They could afford to buy what was cool, and they set the standard for what that was. Boomers had every reason to believe that they would continue into their golden years living the lifestyle they had become accustomed to, thanks to their parents' savings that they would eventually inherit and their own success in business, real estate, and the stock market.

Along came the shock and the reality of the effects of the 2008–2009 recession on everybody's real estate nest eggs, their stock portfolios, and their retirement accounts, bringing about the realization that they hadn't saved enough over the years to make up for it and support themselves in that comfy lifestyle. Boomers kidded about their 401(k)s becoming 201(k)s after taking the hit, but they aren't really laughing. It would probably take more years than they have to play with in the working world to make up for their losses. For seniors, it was just another situation they would have to cope with and they decided to quietly make adjustments to their spending and live with it. For boomers, it went back and forth between being a tragedy and "it's not that bad." That continues today. The drama and the denial are at war with each other.

COPING WITH REALITY

Baby boomers have been treated as special people since they were born. The sheer numbers of them made them appealing to marketers, and they catered to boomers as they had to no generation before them. Boomers knew that whatever they did would influence others. They knew they were important. They got what they wanted in the material sense. Yet being raised by fiscally and morally conservative Great Depression–era parents instilled discipline into them and a sense that they needed to earn their way in the world. Boomers are competitive. They had to be better than the next person to get the job or get chosen for the team. They didn't get little trophies for just showing up, like the Millennials did a couple of generations later.

Boomers mostly have not had to deal with deprivation, at least not nearly as frequently or deeply as seniors before them. So the question becomes, how will they deal with it now? And, how does that affect how you get them into your homes? It may take years to answer those questions. Boomers may play the game of trying to outsmart the next person when it comes to finding the best quality within a lower budget, but it may be a short-lived game. Boomertising's Wallace said, "More than 85 percent of boomers say they'll be more practical and pragmatic in their purchases post-recession. They can say that, but it's in their blood. They need to feel that what they're buying is an indulgence."

You can keep in mind what the boomers—and all of us—have been through. However, before and after the recession, it basically boils down to your going back to the basics. All of the marketing experts agree that for any community, including those of interest to the age 50+ buyer, it is essential for you to

- Know who your customers are
- Know what they want
- Know what they will pay

Market research plays an important role in your being able to answer those questions. It can be formal and extensive or informal and narrow. In a presentation to homebuilders in 2008, ProMatura Group President Margaret Wylde talked about how builders of any size could be "running with the big dogs." Below are some of her suggestions related to research:

1. *Do a feasibility study:* Even if it's just going to a university to get students to help, do something. Or, talk to a firm that knows the industry and if you can't afford a large

study, tell them you want a small one, enough to tell you if there is enough feasibility in your market area to do a project that you have in mind. "Is there opportunity for the product you want to build?" is the question you want to answer.

2. *Learn about the demographics of the area you have in mind:* Who lives there, what characteristics do they have? Can they afford what you want to build?

3. *Learn about the perceptions of residents in your area about the community you have planned:* Focus groups can help too. Find people who are planning to buy a home soon, can afford the home you're going to build, and would at least consider purchasing a home from you. Show them a variety of floor plans and features and get their feedback. "If you hear the same thing from enough people, you might consider changing your floor plan," said Wylde.

4. *Get to know your competitors:* Do mystery shops of your competitors' homes. You can do it yourself if you won't be recognized. Learn all about their product, how they are positioning it, and what they say about you. You can do a reverse address lookup and get a focus group of people who have bought homes from your competitors. Ask them why they bought what they did. You may or may not reveal who the focus group is being conducted for. It's good to know what the competition is doing well and what works, said Wylde.

5. *Be different:* Don't do what everybody else in your area is doing. For example, if everyone is building single-family detached homes, you might want to research row houses. Conduct design charettes. Get your design team and a group of prospective buyers together, spend 3 or 4 hours together, and come up with a streetscape or new design that people might like. Look to do innovation rather than copying what exists now.

6. *Do satisfaction surveys of your own customers:* Ask how you can improve. "Ask the questions you're afraid to ask," said Wylde. About 85 percent of your customers should mark "satisfied" or "very satisfied" on the surveys. You need to know what you can do differently next time. Do a walkthrough with your customers 6 to 12 months after they move in. Ask them what they like and don't like. Do a lost prospect poll. Why didn't they buy after they looked at your homes?

The more you know up front, the fewer missteps you will make.

CAPTURING BOOMERS' INTEREST

Just how do people find out about your project? Even though active adult community homebuyers don't make up a huge chunk of the 50+ market, some related data could be informative. ProMatura Group's 2009 study of active adult homebuyers indicated that 37 percent of buyers learned about the community through the homebuilder's marketing efforts and an additional 20 percent by seeing the community. One-third heard about the community through another person, usually a family member (56 percent) or friend (38 percent). By far the most frequent family member to tell the prospect about the community is the daughter. Of the marketing vehicles, newspaper ads lured 37 percent of the buyers (see Fig. 12-1), the Web 21 percent.

YOU ARE CORDIALLY INVITED TO SPEND A

DAY ON THE BAY

JUNE 4TH & 5TH
Saturday and Sunday, Noon to 4 pm

See

...our 8 styles of townhomes and furnished model overlooking panoramic views.
Preview our spectacular new single level floor throughs.
View the work of contemporary New England artists from The Donovan Gallery.

Taste

...the fine wines of Sakonnet Vineyards.

Hear

...from our design coordinators about the
many exciting options available at The Villages.

Experience

...the pleasure of carefree living at The Villages.

2005 Platinum Award Winner – *Best Active Adult Community in the United States*

*Leave a piece of your registration ticket in each home you tour
to be eligible for any one of our ten special door prizes.*

The Villages
TIVERTON, RHODE ISLAND

For details call **401.624.1300** or visit **www.mounthopebay.com**

Developed by Starwood Tiverton, LLC • Marketed by TCC The Collaborative Companies

Directions: From Boston, Rt. 128 to Rt. 24 South to Tiverton, RI / Exit 5 (Rt. 77 Tiverton/Newport) Right on Rt. 138/Main Rd. for 0.4 mi. to entry on left.

Figure 12-1

Newspaper ad and invitation for a meet-and-greet event with prospects and current residents as "ambassadors." (*Ad courtesy of Carlson Communications.*)

Boomers are far more Web savvy than the previous generation. Nearly three-quarters of people between ages 50 and 64 use the Internet versus less than half of people age 65 and over, according to a December 2008 survey by the Pew Internet & American Life Project. As Internet users increase in numbers as they get older, the percentage of prospects generated by a community's Web site will surely dramatically increase. Recent studies indicated that people 55+ spend more time online than young adults and that conducting Internet searches is an immensely popular and important activity for retired persons, second only to e-mail. And, a study conducted by New York City-based research firm eMarketer led the firm to the conclusion that by 2011 nearly 83 percent of boomers will be Internet users. Therefore, it would pay dividends for you to invest in someone who can guide you in Web site design and navigation, search engine optimization (SEO), and Web site traffic measurement systems.

In general, active adult homebuyers ranked marketing as least effective in influencing their buying decision when compared to model homes, sales representative, clubhouse or other amenities, sales center, and discounts and incentives, which they ranked most effective in that order. Keep in mind, however, that most people are not aware or willing to admit that they are susceptible to marketing messages, so that may not be wholly accurate.

Timeline

Boomers will not be a quick sell, or an easy sell. Most of them probably don't *have* to move, so you have to work on getting them to *want* to move—and make it their idea. Boomers don't like to be told what to do. (Who does?) They'll be looking at lots of other neighborhoods and homes and comparing them all not only to your offerings but also to the home in which they currently live. You know, the home they have lived in for a decade or two or three. The one in the neighborhood they love. The one surrounded by great friends and neighbors. The one with hundreds of great memories. They will forget about the bad times and the pesky neighborhood dog and the drafty bedroom when they consider leaving it behind. "Remember, competition can be remodeling or staying where they are," said Tracy Lux, president of Trace Marketing Inc., Sarasota, Florida.

In ProMatura Group's 2009 study of active adult homebuyers, prospects looked at an average of 4.6 communities before buying and visited the one where they ended up buying an average of 3.8 times. The time from when they decided they wanted to buy a new home and the time they actually purchased, according to the same survey, was 7.8 months.

Consequently, boomers who are playing with the idea of moving will seek lots of information. They, unlike generations before them and after them, will read every word in your brochure, will pore over every page on your Web site, will come back and visit many times the ones they narrow it down to. "The Web is vital because this buyer will actually review every page in detail; this is a reading generation," noted Chuck Covell of Covell Communities, Gaithersburg, Maryland. Home shoppers will already be well informed about your development but they will also ask a lot of questions and they're not afraid to ask the tough ones. So be prepared.

Since they're not in a hurry, you could find yourself getting discouraged and impatient. Will these folks *ever* make up their minds? Will they *ever* commit? Yes, they will, but they will do it their way, which is not the same way the GI generation, the silents, or Gen Xers do it.

Boomers will take their sweet time, which to you may not seem so sweet after awhile. It may take a year for some of them to make up their minds. One or two visits, a couple of conversation and a spin or two around the Web site just will not do it.

The secrets to snagging a boomer are to let them do it their way, let them proceed on their own timeline, and keep them engaged with you the whole time.

Janis Ehlers, president of The Ehlers Group, Ft. Lauderdale, Florida, said, "When prospective buyers may take up to 8 months to make their buying decision, marketing strategies need to adapt to their timeframe. Keeping buyers engaged and returning to the community takes a creative commitment."

Engagement = Relationships

Ehlers recommends using a mix of advertising, publicity, promotions, and e-marketing. E-mails and electronic newsletters are inexpensive ways to stay connected without being too intrusive or annoying, as phone calls or direct mail can be when prospects are not in the mode of buying right then. Consistently sending electronic correspondence keeps you in front of your prospects yet not demanding that they respond to you, so when they are ready, they will remember you and know where to find you.

Being pushy—or even being perceived as being pushy—will only drive away your boomer prospects. Author Schriener looked at one active adult development, walked through the models, and chatted with two sales agents. She told them she was just at the very beginning of the process, that she was in no hurry, that she wasn't even sure what type of house she wanted, that she would be looking at several different house types, and that she wasn't looking to move for several months, probably at least a year. Even after communicating all of that, one of the sales agents called the week after the visit and again a month later, and sent e-mails asking if there were any questions she could answer—in other words, she was pushing Schriener to make a decision, at least in her mind. Schriener was afraid to go back to see if she wanted to put the complex on her short list for fear of being grilled and pressured before she was ready.

The smarter approach is one of engagement. "If the sales staff can learn to be patient and work the relationship, it will pay off," suggested Covell. Working the relationship isn't about keeping a constant one-way flow of information going to the prospect. It involves becoming their go-to person for all issues related to homes, the real estate market, the economy in general, mortgages, selling their current home, and getting their current home ready to sell. If you are helpful and seem to be interested in their welfare even if it means they're not going to move to your property, they will come to you with their questions. That starts a relationship. Then friendly small talk or chit-chat about their kids or grandkids, their latest craft project or auto restoration, their aging parents, the vacation trip they're planning, their upcoming golf game or tennis match, or their favorite sports team's sad defeat last week all turn a good professional contact into a friend. "Guilt!" is a great sales tool, declared Lux. "Tie them to you, and keep in touch all year long so when they're [in the neighborhood] they wouldn't dream of not coming by to say hi."

What excuse do you have to keep in touch if your boomer home shopper is not calling you? Some good options do exist.

There's always some tidbit in the news that you could send them—something you found interesting that you thought they would like to know. You are more likely than they are to run across that type of information since you have access to specialized trade publications and e-mail alerts that your prospects probably don't. By e-mailing your prospects with the info or a link to it, it shows them that you are thinking about them and it keeps you in their current frame of reference.

You could invite them to a social gathering of current residents and other prospects. If your development puts on events as a marketing tool to build relationships, your best prospects are obvious choices for you to invite, but don't neglect the maybes and the probably-nots too. Relationships can turn cold prospects into hot ones very quickly. Even if the relationships that sell them on the value of moving into your community or considering it seriously aren't with you but are with current residents or fellow prospects, you are the one who provided the opportunity to meet and mingle, and they can share their delight with you, all of which only deepens their relationship with you.

Event marketing gives positive reinforcement to people who are uncertain about how they might like living in a community. Deborah Blake, vice president, marketing, for Pulte Homes in Scottsdale, Arizona, said: "People say, 'The pictures in brochures are beautiful, but will I fit in?' So event marketing is a great way for prospects to see if they will fit in."

These events don't have to be elaborate. They can be casual on-site parties with plenty of food and opportunities to mingle with current residents. You might consider establishing an ambassador program, tapping into happy, enthusiastic current residents for these events, people who would carry a positive image and appealing message to prospects. Real estate laws now prohibit compensating non–real estate professionals monetarily, but you can lavish praise and gratitude on them, raffle off a large-screen television to one of them, make a donation to the home owners association for an improvement they haven't been able to swing, or give them some other reward. Covell suggested setting up the ambassadors ahead of time, matching them up with prospects with similar or compatible interests and backgrounds, and letting them get into "me too" conversations so they don't feel alone. "The key is to make the emotional connection," emphasized Rich Carlson, who heads Carlson Communications in Northborough, Massachusetts.

MESSAGING TO THE "WHO, ME? I'LL NEVER GET OLD!" GENERATION

Boomers will never get old—just ask them! A majority of them will still not be describing themselves as old when they're well into their seventies, maybe even their eighties. Take universal design, for example. "Even though [aging boomers] have certain requirements for their aging bodies, the way to appeal to them isn't on that basis and never has been," said Boomertising's Wallace (who hates the term "aging boomers"). But boomers do like for life to be as easy and convenient as possible and universal design (UD) provides that. Boomers also like to be in on the latest and greatest and if UD is trendy, they are likely to be open to it. And, understanding that UD features hike up the resale value and increase the number of prospects if and when they decide to sell their home in the future increases their receptivity.

Remember, also, that boomers are very likely to stay active as long as they can, and that doesn't just mean playing golf. Many are into very physically demanding or risky sports, such as rock climbing, mountain climbing or steep hiking, race car driving, motorcycling, snowboarding, snow skiing, and water skiing in their forties, fifties, and sixties and plan to continue into their seventies and eighties. In recognition of the boomers' bent to be outrageously active, AARP developed motorcycle insurance, so this is not a small group to consider. Increased strenuous activity brings increased likelihood of injury: muscle tears; cartilage and tendon injuries; sprained or broken limbs; hip, knee, and shoulder replacements; and possibly even paralysis or blindness. It's quite likely that boomers will need some kind of accommodation for physical limitations, at least temporarily at least once, and when they do, those accommodations better be in place. If they're not, installation of the necessary elements can be disruptive and expensive, if they're possible at all, and recovery can be severely hampered. Of course, you don't want to bring up these disturbing possibilities to your homebuyers but you can incorporate what they might need into your design.

Boomers Don't Want to Hear ...

- That they're getting old
- The words "senior," "elderly," "old," "crippled," "handicapped," "disabled"
- Music from the 1960s
- Anything about shuffleboard or bingo

Boomers Do Want to Hear ...

- That they are forever young, healthy, and vital
- That their home will be easy to live in
- That they can be social in their home
- That their home will have benefits galore

For example, talk about "the next phase" of their lives instead of "retirement." Use "convenience" and "flexibility" when describing universal design; don't talk about how it will help them cope with an infirmity. Remind them that going green will save them money, not just that it's the right thing to do. Even if they believe in considering the environment, it will be the personal benefits to them that will help tip them over to decide to move. Do not insult them by playing music supposedly from their era of the 1960s and 1970s. When a marketer tries to lure boomers with music from 30 or 40 years ago, it's usually because someone in their twenties or thirties thought it was a good idea. As far as boomers are concerned, their era is now. The past is the past and they are firmly in the present, if not the future.

IT'S ALL ABOUT LIFESTYLE

Just like the secret to real estate success is location, location, location, the secret to becoming a successful builder for boomers is lifestyle, lifestyle, lifestyle. The authors may repeat themselves several times in this book when it comes to emphasizing lifestyles, but it's worth repeating. Sometimes the lifestyle that would be attractive to a segment of boomers is already built in to the project; it just needs a marketing message that's targeted to boomers to make them aware of it.

Artisan Lofts, the loft project built by Artisan Homes in Phoenix that is mentioned in Chap. 7 is an example. The builder envisioned that the development would attract young urban dwellers looking to live in an urban setting, close to restaurants, conveniences, and "the action," Artisan developers were astonished to discover that as many as half of their buyers were singles or couples over 50 years old. By emphasizing the lifestyle in their marketing message, that percentage increased. Before, they did not mention empty nesters, one-level living, or anything else that directed the message to the 50+ household. After seeing the potential benefits of the older market, they crafted their message to the lifestyle features that the 50+ would enjoy: the large pool and athletic center, the downtown location, the one-story living units (along with some two-story units) and the community gathering spaces and retail outlets conveniently located in the building. Then it became a place that empty nesters could see themselves in. It was the *lifestyle* that appealed to both the young singles as well as a significant segment of 50+ buyers.

MARKETING IN THE NEW AGE

Forget One-Way Communication

The days of blasting out a message to customers en masse are over. People just won't stand for it, especially boomers, who have always wanted to weigh in on every issue and have their opinions mean something. The reality television shows "got it" very early on. They realized that by involving viewers, not just the audience in the studio, and asking them to vote to help determine who got to stay and who had to go, they could get more buy-in to whatever premise they had created. Voters had a sense of ownership and kept coming back. The concept of shared ownership—or at least making people *feel* as if they have a say in what goes on—now is ubiquitous in television and other media. That carries over to marketing in general.

Social Networking

The Web has taken interactivity to great heights, especially since social networking has taken hold and started to become influential. Social networking sites—also known as social media—are where people go to connect, share, collaborate, and exchange and discuss information. Social networking is usually a gathering of individuals with something in common, which may be, to name a few, an interest in politics, hobbies, sports, foreign languages, music, television programs, health issues, or lifestyle similarities such as raising teenagers or taking care of elderly parents. LinkedIn, Facebook, Twitter, MySpace, FriendFeed, Classmates.com, Meetup, and Gather are some general social networking sites where members can "meet" with others in narrower interest groups that the members themselves create. There are also hundreds of special-interest online communities where people gather, and more pop up daily. These sites are getting to be *de rigueur* for companies and communities in the residential development, architecture, and building business, as well as every other kind of business-to-business and business-to-consumer organization. Like the Web evolved into a place where everyone had to be a decade and a half ago, social media sites are next.

Marketers can join these social networking sites in two ways. One, they can join as a professional entity—community, parent company, management company, etc.—using the company logo or some other professional image as their profile "avatar." People expect a professional entity to do a certain amount of promotion of their own products on social networking sites so they shouldn't be too put off if you do that. Or, an organization may want to have individuals join the online networking circus and be a little more personal in their interactions. The individuals can let people know in their profiles where they work but also relate to fellow social networkers in ways that make them human and provide mutually shared experiences and opinions to create bonds. The ideal is to have both an organizational presence and also at least one individual participant, preferably more. Then the company or community can reveal some piece of information or promotion from that account and the person who works for the organization can relay the same information as coming from the company, so you are double covered. The relationships built by the individuals give them credibility and hence the messages carry more weight than if they just came from the community. The individuals who participate have to be personally interested in the experience of social networking or it won't work, because social networkers can easily tell who is there to sell them something and who is there to form relationships.

Boomers are natural networkers in real life. They have established networks of contacts, colleagues, and friends wherever they've gone all of their lives. So taking that propensity to the Web is not a great stretch, which is why statistics show that boomers are getting to be as big of a presence on social networking sites as their children and grandchildren.

Specifically, some of the social networking sites where you can build a presence are

- *Blogs:* Originally named such as a contraction of "Web" and "log," a blog as posted by an organization or corporate entity is halfway between an official company message and a personal insider's look at the message and the company behind it for the benefit of the reader. A blog is like a column written in a newspaper by someone whose observations, facts, opinions, and humor you come to like, respect, and trust, only it's online. Blog entries may be long or short, mostly facts or mostly analysis, and serious or filled with amusement or humor. What matters if you want to engage people is that blog entries are frequent, engage the reader (i.e., don't preach), and invite comments.

Blog entries ideally are meaty and have something in them that people can either identify with, argue with, cheer you on about, or laugh with you about. Recruit some residents to make comments to get the ball rolling and, if appropriate, respond to the comments to give the blog immediacy and personality.

A blog is a great place to launch a test balloon for an action you're contemplating. *We're thinking about turning one of our card rooms in the clubhouse into a conference room so our residents who work at home can hold meetings there instead of having to use their homes or go somewhere else. What do you think?* A blog is where you can give readers an inside look at some goings-on, not unlike gossip. Who doesn't like to have the inside scoop? *The mayor stopped by yesterday for lunch and a tour with his mother, one of our newest residents. A small crowd formed and he found himself shaking hands, petting dogs, and saying maybe to invitations to play pickle ball.* And, a blog is a good place to just inform people about the status of the development. *We are so excited to be able to tell you that our second outdoor pool and hot tub will be open in time for next summer!*

A blog should be written casually and not like a press release or an advertisement. It should be light and personal. Look at what your competitors are doing and organizations in other industries. You'll get the hang of it quickly and be able to adapt it to your own community.

- *LinkedIn:* This is a place for professional connections and discussions. Individuals can join and get linked with other individuals. They may position themselves as experts on different subjects in an "answers" section where LinkedIn members ask questions and anyone with expertise in that area may answer. Anyone may start a professional group on LinkedIn for their organization and manage discussions within that.

- *Facebook:* Hundreds of millions of people have "friends" on Facebook. What started out as a way to keep up with personal friends has become more and more a place of great interest and benefit to the business world. Business colleagues and acquaintances who had rare contact in person now follow each other's news streams on a daily basis, learning sometimes intensely personal things about them. As a smart marketer, you can use this to your advantage, sharing links to articles of interest to people, keeping people up to date on your business activities, and sprinkling in some human interest info to let people know you're human. You may also start a group on Facebook and ask people to be "fans." Growth of those groups tends to be "viral," in other words spreading by word of mouth from one person to another to another, and you can find yourself with a healthy group of fans in a short amount of time. The trick, then, is to keep the group current so fans will hear from you from time to time with information of interest to them.

- *Twitter:* Quirky Twitter is based on what is known as microblogging. The idea originally was similar to blogging but in "tweets" of 140 characters at a time. It caught on and became a true social networking opportunity for millions of people. You sign up, begin "following" people, which means their tweets come into your Twitter home page, and gather followers yourself, other "tweeple" who follow your tweets as you "twitter" or tweet them, and they flow in to their Twitter home page. Your community can have a presence on Twitter and so can individuals. Once you have a following built up, some of your followers choose to "retweet" some of your good tweets, which then go to your followers' followers. Confusing? You have to try it and spend some time on Twitter to see how it works. You can announce events, contests, price changes, and other special opportunities. Or you can tweet about things that are happening in your community to build a positive image. People who are interested in a community like yours can discover it on Twitter with a little diligence on your part.

- *YouTube:* If you take the time to make short videos of activities or events at your community, or even a tour of the homes and amenities, you can post it on YouTube and send the link to people who might be interested in seeing it. You can post your latest television commercial if you advertise that way. It's a shortcut, a cheap way to enable people to see what you're up to or deliver a message to them.

Social networking is changing so fast that the moment the authors type this, it is out of date. There are plenty of resources on the Web that can tell you what social network sites are hot at the moment—or some other innovative online idea that may pop up—and how to use them to your advantage. The authors just wanted to make you aware of some of the social networking sites and help you see that they can help you establish relationships and deliver your message in a new, nontraditional way.

REFERENCES

1. "50+ Builder Forum: Surviving the Slowdown, Planning Ahead for the Upturn." National Association of Home Builders Webinar, December 15, 2009,
2. Blake, Deborah, Rich Carlson, and Chuck Covell, "Getting Hesitant Boomers to Buy Now," National Association of Home Builders audio seminar, August 12, 2008.
3. Childress, Donna, "Are You Tech Savvy Enough for Tomorrow's Seniors?" *Innovations*, National Council on Aging, Summer 2009, pp. 20–22.
4. Connell, Regina M., Shuck, Barbara D., Thatch, Marion, *White Paper: Social Networking for Competitive Advantage,* Society for Marketing Professional Services Foundation, Alexandria, VA, July 2009.
5. Green, Brent, *Marketing to Leading-Edge Baby Boomers*, Paramount Market Publishing Inc., Ithaca, NY, 2005.
6. Jones, Sydney and Susannah Fox, "Generations Online in 2009," The Pew Internet & American Life Project, Washington, D.C., January 2009.
7. Kleyman, Paul, Journalists Exchange on Aging Survey on Style, Age Beat: The Newsletter of the Journalists Exchange on Aging (part of American Society on Aging), No. 24, Special Report, Summer 2007.
8. Phillips, Lisa E., "Boomers Online: Attitude is Everything," eMarketer, New York, NY, December 2008.
9. Wylde, Margaret, *Boomers on the Horizon: Housing Preferences of the 55+ Market*, BuilderBooks (National Association of Home Builders), Washington, D.C., 2002.
10. Wylde, Margaret, "Active Adult Housing Prospects," National Association of Home Builders Webinar, September 2009.
11. Wylde, Margaret A., *Right House, Right Place, Right Time: Community and Lifestyle Preferences of the 45+ Housing Market*, BuilderBooks (National Association of Home Builders), Washington, D.C., 2008.
12. Wylde, Margaret and Janis R. Ehlers, "Run with the Big Dogs," presentation, International Builders' Show, Orlando, FL, January 2008.

A Cohousing

Charles Durrett was generous enough to, at the authors' request, write a piece exclusively for this book describing cohousing. Durrett, along with his business partner and wife Kathryn McCamant, wrote the original book "Cohousing: A Contemporary Approach to Housing Ourselves", which was first published in 1988 by Ten Speed Press. They have since gone on to write several books and handbooks on cohousing. The latest by Durrett, "The Senior Cohousing Handbook, 2nd Edition: A Community Approach to Independent Living," was published in 2009. As a result of their efforts, cohousing was introduced to the United States. The entire text of what Durrett wrote for this book is included below. The authors believe that cohousing, while not specifically for the 50+ market, is poised for rapid growth for the boomer market in the very near future. Cohousing communities are small, which is perfect for custom building companies and other smaller volume builders. These communities can be tailored for any 50+ niche, no matter how small. Following the initial section by Charles Durrett is a case study written by the authors on a builder and developer, Wonderland Hill, which has been successfully developing and building cohousing communities for 30 years.

Senior Cohousing: The Village Solution

(by Charles Durrett and Marysia Miernowska)

Introducing Cohousing

Cohousing is a grassroots movement that began in Europe in the early 1970s and grew directly out of people's dissatisfaction with existing housing choices. People were tired of the isolation and impracticality of conventional single-family houses and apartment living, and desired a more sociable home setting that had less of an impact on the environment. The first cohousing communities were formed to combine the anatomy of private dwellings with the advantages of community living. The developments varied in size, financing method, and ownership structure, but shared a consistent theme on how people could cooperate in a residential environment to create a stronger sense of community and to share common facilities.

In this appendix we will examine how the cohousing model has adapted to meet the needs and expectations of seniors and active adults in the United States. We will explain the characteristics of this neighborhood model, present design guidelines, and suggest how it can be used as a noninstitutional alternative to aging in isolation. We will demonstrate how senior cohousing has become a socially and environmentally sustainable solution, allowing successful aging in place. We will explain the process that you can use to explore issues of aging in place and to determine if community is for you. Finally, we will reveal how to use senior cohousing to bring you a fulfilling, community-oriented aging experience with a high quality of life.

The Design Guidelines and Characteristics of Cohousing

Cohousing communities typically cluster private residences around common spaces and facilities, often preserving open space as well. The common house is the heart of the community, and it serves as a gathering place for group activities, meal preparation, common dining, and meetings. The size of the common house depends on the people in the community, the frequency of meals to be shared, and the budget available. Common houses may include guest rooms, multipurpose/hobby space, a meditation or prayer room, and a fitness center. It's entirely up to the group to decide on the features to include.

Parking and roads are typically located on the perimeter of the property in order to leave the center of the community as a pedestrian environment. People strolling or rolling by can stop and chat with neighbors rather than enter their homes by car through the garage, as in typical neighborhoods. The common house often acts as a "gateway" from the parking area to the community, thus providing another opportunity to interact with neighbors getting mail, or while seeing what is cooking for the evening's common dinner.

Pedestrian streets or a series of courtyards provide the primary circulation to and from parking.

Along the way, you will find "gathering nodes," which can be as simple as a picnic table, surrounded by flowers, where neighbors can meet to have breakfast on a Saturday morning or share a beer in the afternoon. These gathering nodes are usually visible from about four to five surrounding homes.

Indeed, visibility is reflected in the site lines of the site plan, as well as in the layout of the private homes. Unlike typical suburban homes, the kitchen faces the community, thus providing a view from the kitchen sink. The living room, on the other hand, is located to the rear of the home, opening to a private backyard.

The front porch becomes as lively as in traditional neighborhoods; it's a great place for people-watching, thereby facilitating personal interaction with people outside the home.

Cohousing communities may include retail space, as at Frog Song in Cotati, California, and affordable homes may be mixed with market-rate homes, as at Silver Sage in

Boulder, Colorado. The homes can be lofts in renovated industrial areas; they can be townhouses, condominiums, single-family detached; and they can include existing and new homes within an established neighborhood. Any building form will do as long as the community design incorporates the essential design elements: common areas both indoors and outdoors; a pedestrian environment leading to entries to the homes; socially and environmentally sustainable design; and a common house where meals and other events can be shared.

Indeed, the four main characteristics of cohousing are extensive common facilities, design that facilitates community, a participatory development process, and complete management by the residents. In a series of meetings, future residents work with the architect to design a neighborhood that reflects their values and goals.

Once the neighborhood is built, residents remain responsible for the ongoing management of the project. Major decisions are made at common meetings, and responsibilities are divided among committees that honor a decision-making process based on consensus.

The Senior Dilemma

Just as people were dissatisfied with the impracticality of conventional housing for their families, seniors and aging adults have also found their needs are not being met. The typical suburb, where the automobile is a de facto expansion of the single-family house, people are essentially required to drive to conduct business or join in social opportunities. As we get older and our bodies and minds age, the activities we once took for granted aren't so easy anymore: the house becomes too big to maintain; a visit to the grocery store or doctor's office becomes a major expedition; and if we can't drive to a friend's house for a visit, then we stay at home alone all day. Isolation is a serious fear; so is being a burden to one's children, or having one's responsibilities and freedoms taken away in an institution. Of course many, if not most, seniors recognize the need to effectively take control of their own housing situation as they age. They dream of living in an affordable, safe, readily accessible neighborhood where people of all ages know and help each other. But then what? How many of those kinds of housing choices actually exist?

The modern single-family detached home, which constitutes about 70 percent of the American housing stock, is designed for the mythical nuclear family consisting of a working father, a stay-at-home mother, and two to four children. Today, less than one-quarter of the American population lives in such households. Almost one-quarter of the population lives alone, and this proportion is increasing as the number of Americans over the age of 60 increases. At the same time, the surge in housing costs and the increasing mobility of the population combine to break down traditional community ties.

Currently, seniors (those over 55) represent 12.4 percent of the American population which, with the swell of post–World War II baby boomers entering the senior ranks, will increase to 20 percent by the year 2030. Clearly action must be taken, and quickly, to correct these household and community shortcomings. But what can be done, and by whom? How can we better house ourselves as we age?

Senior Cohousing: A Proven Approach—A New Application

Usually limited to those 55 years of age or over, senior cohousing takes the concepts of cohousing and modifies them according to the specific needs of seniors. The result is a cozy little village that invites involvement, cooperation, sharing, and friendship—a recreation of earlier times when community participation was viewed as an essential part of social, mental, and physical health.

Exploring the social, mental, and physical issues of aging was the main purpose of senior gatherings organized in Denmark. Faced with similar challenges of the single-family house, seniors began getting together and discussing the issues of aging in place. Now called "Study Group I," these workshops were a great success. Seniors finally had an opportunity to face the challenges of aging with their peers. As seniors supported and inspired each other through the process of discussing these issues, they became empowered in their new social roles and developed relationships that created a strong foundation for continued growth in community. Time and time again, at the completion of the workshop, a group of empowered seniors concluded that aging in community would best help them achieve the quality of life that they desired.

Top 10 Reasons People Want to Live in Cohousing

1. *Save money:* The average house in cohousing is less than typical houses and day-to-day costs are less (shared lawn mowers, guest rooms, workshop, and much, much more).

2. *Child rearing:* Some people live in cohousing so that they can have only one child, and don't have to worry about driving their child to play dates each weekend. In any case, children have a lot of playmates in cohousing.

3. *Takes pressure off of the family:* There are others to help, to talk to, consider, and to go to the movies with.

4. *Lifelong learning and cultural enrichment:* There are others around who know how to do things and they teach each other. Hardly anyone watches TV in cohousing. They play music, or talk about music, or the play, or the movie, or the art.

5. *Living lighter on the planet:* For example, our electric bill last year was $83.84, our heating was about $20 per month, we drove 25 percent less (much of our "community" lived in the neighborhood), and we educated each other about how not to need air-conditioning in our hot summers and about other energy-saving devices.

6. *More practical and more convenient:* It's more practical to have others around to lend a hand when you need one. Our neighbor on occasion picks up something at the grocery store, gives a neighbor a ride to the store, or an elderly member to the pharmacy, or carpooling the kids to school, and much more.

7. *Live more intergenerationally:* In Nevada City Cohousing there are 37 kids and 21 folks above 55. In typical suburbs it is much more difficult for kids and seniors to spend time together and much less likely.

8. *Live more healthfully:* Others around often demonstrate good exercise habits, good eating habits, and no one denies staying connected, that is, community in terms of healthful living.

9. *More interesting:* Dinner at the common house, on occasion, leads to provocative conversation.

10. *More fun:* Whether it's folks playing music on the common terrace or swimming races with the kids in the pool—sometimes cohousing feels like summer camp.

These people, the ones who first envisioned senior cohousing, were ordinary citizens who, deeply passionate about their beliefs, decided to help make the dream of socially and environmentally healthy senior neighborhoods a reality. Along the way, they learned many invaluable lessons for creating, building, and living in a seniors-only cohousing community. In a nutshell, senior cohousing is similar to the multigenerational cohousing model, with the following modifications:

- Careful agreement among residents about co-care and its limits
- Design considerations appropriate for seniors
- Size limitations (a maximum of 30 living units, usually 15 to 25)
- Senior-specific methods for creating the community

Co-care

Co-care is a unique quality of senior cohousing. Its presence allows seniors more flexibility in terms of the care they receive and give. It has immense benefits, especially in terms of the quality and economics of care. Co-care exists because of the relationships and the proximity that exist in cohousing neighborhoods.

"Living in senior cohousing has given both my wife and myself a quality of life that we could not have received anywhere else. Being legally blind, my wife almost always needs somebody around. Being so close to our good friends and neighbors, she remains very socially engaged, and I can have time for my own personal activities without worrying about her."

Basic co-care issues are resolved far before move-in, during the Study Group I workshop, where future residents discuss issues of aging in place. Often the constituencies that are formed are extremely brief. For example: "If you haven't raised your blinds by 10 a.m. someone will knock on your door." A group will undoubtedly want to establish boundaries; after all, most residents just want to be neighbors, not health aides. Nonetheless, cohousing facilitates helping a neighbor, where otherwise help must be hired. Picking up someone's medicine while you're picking up your own is just not that big a deal. Neither is having tea with a sick friend, or bringing them the paper.

Co-care basically falls into three categories:

- That which you have agreed to do, which might include such things as bringing dinner to people when they can't make it to the common dinner.
- That which you agreed you would not do, for example, bathe someone.
- That which you never imagined you would do before you cared about this person, but it turned out to be easy and he's just next door so it's convenient enough. (I've noticed cohousing to be readily available karma—you do things for others and before you know it, someone's doing something for you.)

Senior cohousing also facilitates on-site professional care in a manner that is both more economical and more pleasant for everyone. Often, residents will choose to build an affordable apartment above their common house for an onsite nurse. The costs of this luxury are split among the residents, thus making it more affordable. What makes these caregivers unique is that they are hired and fired by residents. Furthermore, as the nurse takes care of the older residents, she gets to know younger residents in a way that will make her more accountable to them when it is their turn to be cared for. This happens mostly because she knows them as real people, not just clients; not just as someone that you have to care for, but someone with whom you have a lot in common.

Senior Cohousing Case Studies

1. Munksogard: Strong Roots in Fertile Soil   The Danes, like Americans, value the ideals of individualism and self-sufficiency. Therefore, a look into a Danish senior cohousing community, Munksogard, helps us imagine how community can enhance the experience of aging, while maintaining privacy and independence. The community is very successful; as of the time of writing, none of the residents have moved out and over 80 people are on the waiting list to move in.

The goal of Munksogard was first established in July 1995, by a group of environmentalists from Copenhagen. In short, it would become a modern version of the traditional village, one that pointed toward the future and could inspire others. Additionally, the community was to sustain both environmental and social ecologies.

Cars are not permitted beyond the parking lot, but there are plenty of wheelbarrows standing around for people to carry their groceries from their cars to their home. The seniors enjoy a variety of common activities such as playing bridge, singing in a choir, picnicking, and sharing common dinner in the common house 3 or 4 times a week. The residents also enjoy working together in the vegetable garden. Since moving in, they have built a workshop, a storage shed, a greenhouse, a clothes drying shed, and a henhouse. While the younger seniors painted the houses, the older residents cooked and made coffee for the group.

The unique social advantages of senior cohousing communities can be seen in the activities that occur and in the stories of the residents. Naturally, as cohousing communities are entirely resident-managed, it is the residents who decide what common activities will be organized. Nonetheless, the unique joy of these communities comes from the abundance of spontaneous activities facilitated by the environment and the relationships formed.

The residents chose to leave a large part of the site undeveloped, partly as a recreational area and partly as a vegetable garden. With the homes clustered, the seniors can all easily see if anything exciting is happening in the courtyard. Walking home, they can join their friends for a cup of tea on their common terrace. If they choose to sew and would like company, all they need to do is go into the crafts room in the common house, and minutes later others have gathered to join. On the other hand, if they would like privacy, the home and the backyard are acknowledged as sacred places for that purpose. This is the choice that cohousing provides: you can have as much privacy as you want, and as much community as you want. Another clue why senior cohousing works for both the Danes and Americans; both cultures like freedom of choice.

In this regard, Munksogard is profoundly Danish; the residents aren't so much living in a mutually dependent relationship, as they are engaged in a mutually beneficial partnership. And it's even more than that; these seniors have created a community, a village where friendships run as deep as family; a world where each individual's social, physical, and emotional well-being is sustained one dinner at a time, one project at a time, every day, in the company of others. Concluded one resident, Dejgaard, "We look after each other. We eat together, talk together, and are interested in each other."

Rich, another resident, walked into the Common House and took the gong off the wall. "Have you seen our round table?" he asked. "We designed it ourselves." He smiled and began walking up and down the courtyard, summoning his neighbors to their weekly meeting.

It's pretty impressive to think that 12 people can plan something together, in consensus. What is even more astounding is that these 80-year-olds brainstormed a housing solution that they agreed would prevent both isolation and institution. They shared their concerns with friends and began searching for a senior housing model that would address their needs and aspirations.

"The children were long gone and my husband and I were alone in that big house. The lawn was just too much to take care of, and I easily imagined feeling isolated in that house. So a few of us signed up for the University retirement community. They had even agreed to put us near each other. But that development was so large . . . " Residents did not want to feel anonymous in a sea of homes. "There had to be a better way. We were determined to discover it."

"Ellen [another resident] was the one who really thought about who would take care of us when it got to that point. So we listened to her concerns and we started talking about it." None of the friends knew the perfect solution for their retired years, but their friendships were strong so they agreed to explore the issues of aging together. Ellen made invitations for a gathering at her home and she invited her women's group and their husbands, and some other folks from their church. For that first meeting, her living room was full.

The group got together regularly and shared their concerns about the realities of aging. Many were worried about becoming isolated. Even though they were involved in many social activities, they were aware that as they aged, driving everywhere would become less comfortable and maybe impossible. The question became obvious: were we going to just wait for this to happen and then feel sorry for ourselves? Another concern was providing care for a spouse who needed constant assistance. To understand what it meant to care for an elder, they invited a woman from hospice, who went through all of details, explaining the reality of care. They concluded that in proximity, they would avoid isolation and could better support and care for one another. Although options such as the University retirement development could give them the proximity they desired, they knew it was not the solution they were looking for. In the long run, they did not want to live in an institution and they wanted to stay independent, for as long as possible. They continued exploring their options.

"At one point we even considered buying bungalows together. We also looked into renovating a huge house. We had various ideas, and we invited specialists to come and teach us about our different options. We had someone come to talk to us about co-ops. But we decided that a co-op was not what we were looking for either; we wanted more autonomy, and we wanted our children to inherit our homes."

Senior cohousing gave them both the independence and the proximate community they needed to age in place successfully. Having gone through the process of choosing the best housing solution, they embarked on a journey as developers, with a consultant and a hired architect. Four years later their neighborhood was built. "It took a lot longer than we thought," said one resident. "Had we used a more efficient process, we probably would have saved $40,000 to $50,000 on each of our houses."

The example of Glacier Circle is unique, and it provides great lessons for other senior cohousing groups. This group of friends was the first group of seniors in the United States to build their own community. They accomplished it with the average resident age being 80 years old. Their strong friendships and faith in finding a better housing solution carried them through to the successful completion of their community. Nonetheless, what they came to organically is possible to accomplish in a more economical and timely fashion using the process suggested by cohousing professionals.

Residents of Glacier Circle said: "We really wish we would have started this earlier. We probably were not ready at the time, but it would have been great to start aging in community by the age of 60 or 65. If we could share one piece of advice to those considering senior cohousing, that would be it: Just do it!"

The Glacier Circle community is a living example of the vitality that senior cohousing supports. And the lively and caring residents have so much to applaud; after all, they figured out the Study Group I process of exploring issues of aging in place, all on their own. Just as it is intended, this process of exploration and discovery led them to establish a strong foundation upon which they built a thriving community.

2. Silver Sage: A Village for Seniors   More and more cohousing communities are faced with the question of how to accommodate their senior residents, as they grow older and their needs and physical abilities change. Some of the first cohousers to actively address the issue were the residents of Nyland Cohousing in Boulder, Colorado, 10 years after their own community had been built. Their first idea was to extend their community with more units especially designed for the needs and preferences of their senior residents. They also considered subdividing their 42-acre site to create a second cohousing community for seniors, but unfortunately that didn't work out. In order to find out more about the possibilities of making senior cohousing, they hosted two public presentations about Danish senior cohousing that the Cohousing Company conducted during two winter holidays.

The Nyland residents found more interested seniors, and in 2003, they had gathered a group. Their vision was a small-scale participatory cohousing community where everybody knows each other and amenities are close at hand. The developer who had built Nyland, Wonderland Hill Development (see "Builder's Perspective on Cohousing" later in the chapter for more on the builder's perspective), then proposed a 1-acre site in the new-urban Holiday neighborhood 2 miles north of downtown Boulder. Holiday, a new residential neighborhood, includes community gardens, a 1-acre park, bike trails, artists' studios, pedestrian walkways, and a projected Boulder Public Library Branch. The prospective site was located just across the street from Wild Sage—an existing intergenerational cohousing community built in 2004.

Working with a local nonprofit housing advocate, Wonderland Hill Development put together a varied income project and worked though all but one of the steps for creating senior cohousing: Feasibility Phase, and Study Groups I and III. Our firm, The Cohousing Company, was hired to work with the group through the participatory design and development process, Study Group II.

The visions and values that the Silver Sage residents established were

1. Nourish the body and soul with good food, good health, and good company.
2. Live mindfully in community, encouraging wisdom, compassion, and interpersonal growth.
3. Experience stylish, thoughtfully designed interior living.
4. Share inviting outdoor spaces such as gardens, courtyards, decks, patios, and views of the Flatirons.
5. Enjoy North Boulder's urban options, including hiking, biking, cafes, and public transportation.

Once the group completed the design workshop, their values and the activities that they wanted to accommodate were clear. The project was designed in our office and the construction documents were prepared by a Boulder architect, Bryan Bowen.

The residents helped design the common house, which has a large kitchen, a dining area, intimate sitting parlors, guest rooms, and crafts and performance areas. A variety of common activities take place there, including common dinners, speakers, films, house concerts, reading groups, fitness classes, and more. Silver Sage residents share an interest in spirituality, although this is not the emphasis of the community's values.

Creating a Senior Cohousing Community

The recommended process for a successful community is referred to as "Nielsen's Five Phases." Each phase presents its own set of challenges and rewards, as a collection of individuals becomes a community. These explanations are borne out by more than two decades of creating successful cohousing communities. The phases are

1. *Feasibility Phase:* Do we have a project?
2. *Information Phase:* Spreading the word.
3. *Study Group I:* Building the foundations of community.
4. *Study Group II:* The participatory design and development process (3 months).
5. *Study Group III:* After move in policies.

Note: For more detailed information, refer to the any of Durrett's books.

Conclusion

The impact a community can have on an individual is positive, powerful, and well documented. Here's what one senior cohousing resident had to say about her experience: "I wake up in the morning and look forward to a visit from at least one friendly neighbor. There's an element of surprise and interaction. Here, I feel a strong sense of belonging and therefore a stronger sense of caring and being cared about."

The secret is simple; people thrive in a community that interacts and cares. The town of Rosetta, Pennsylvania, has been the subject of numerous studies to discover why such a high percentage of its residents live to be well over 90—an age significantly above average. Surveys show the people here smoke at an average rate; they eat red meat, and drink wine. Such studies repeatedly show that the number one reason for their longevity is the town's heightened sense of community.

After all, a home is more than a roof over one's head or a financial investment. It affects the quality of a person's general well-being: one's confidence, relationships, and even one's health. It can provide a sense of security and comfort, or elicit feelings of frustration, loneliness, and fear. A woman who worries about when to shop for groceries and get dinner on the table while taking care of an ailing spouse, is often unable to concentrate on a job or reserve time to spend with friends or other family members, let alone take time for herself. This aspect of housing cannot be measured by cost, rates of return, or other traditional real estate assessments. A more important concern for senior housing should be the people themselves and the quality of their lives.

The men and women living in senior cohousing communities are perhaps the most honest, clear-eyed, and jovial people I have ever encountered. They completely accept the fact that they are aging. They admit they can't do everything they once did. That's life. But acknowledging this basic truth does not mean they are fatalistic. Rather, they have taken charge of their remaining years with the expressed intent of achieving the highest quality life possible, for as long as possible. For them, this meant choosing to build their own community where they live among people with whom they share a common bond of generation, circumstance, and intention. And they have a great time doing it.

> Hey, we're getting older, and we're going to make the most of it. We've had a lot of experiences, and now we're going to have some more.
>
> Charles Durrett

The Builder's Perspective on Cohousing

The following is a case study of a builder who has made a business of developing and building cohousing communities.

Wonderland Hill's Silver Sage, Boulder, Colorado

Jim Leach founded Wonderland Hill Development Company 30 years ago, along with his homebuilding company at the time, Wonderland Homes. He later sold the home-building arm to concentrate on development.

Wonderland Hill Development Company has completed 16 cohousing communities in Colorado, New Mexico, and California, and has four more in various stages of development. Its earlier communities were constructed by general contractors but Wonderland has now started a new homebuilding company. Leach found that it was taking as much management time to work with the contractors as it would to do the construction in-house. Moreover, a homebuilding company would create an additional profit center.

The Wild Sage community in Boulder, Colorado, is a family-oriented community encompassing all ages, including several families with children. There are 34 homes in Wild Sage on 16 acres, a density of 21 homes per acre. Homes are grouped in two- and three-story townhomes. Nine permanently affordable homes and flats and four "Habitat for Humanity" homes are included in the mix.

Wild Sage is directly across the street from Wonderland Hill's first 50+ cohousing community, Silver Sage, in the same "Holiday" neighborhood of North Boulder. Silver Sage has 16 homes on 1 acre, a density of 16 homes per acre, obviously; six permanently affordable homes are included. Silver Sage Village has a two-story common house with one elevator serving all second-floor homes via a second-level walkway on the sunny south side of the central courtyard. Silver Sage Village incorporates the latest in energy-efficient design to lower energy costs, as is typical in all of Wonderland Hill Development's communities. One of its upcoming developments, Solar Row, will be a "zero energy ready" row-home development in a contemporary style.

According to Leach, Wonderland uses various architects for design work but the firm has formed a special relationship with architects McCamant and Durrett for future California developments. McCamant and Durrett designed Silver Sage in association with local architect Bryan Bowen.

Wonderland Hill is the largest developer/builder of cohousing in the United States, and business is growing. Leach said he doesn't understand fully why other builders haven't entered the market, unless it just looks too hard to learn for a relatively small niche market. Leach and his company, which includes his daughter, Shari Leach, provide a full palette of services to cohousing groups. They conduct workshops to teach people how to form a group; facilitate group formation; provide project management, conflict resolution, and marketing; and offer legal and financial support to newly formed groups. Shari Leach has written a book on the process, *Head, Heart and Hands*, which the company uses to guide people through the process.

Jim Leach insists that his company is no different from a conventional homebuilding company. Owners and buyers must invest in the community at the beginning and participate throughout the process. He is working to train new groups to handle their own management after construction with a self-managed homeowner association.

Every development done by Wonderland Hill is sustainable to the degree possible within the constraints of budgets and technology. On the wall of Wonderland Development's office reception area is the company motto: "Community is the secret ingredient in sustainability." That motto emphasizes the fact that green/sustainable technology is worthless in the hands of people who don't care or are untrained in its operation. Cohousing devotees are generally people who are concerned about the environment and are willing to learn how to live sustainably.

Wonderland and others in the cohousing business have found that cohousing buyers:

- Are spiritual people
- Believe in community
- Care about the environment
- Are all ages
- Are willing to help each other
- Are accepting of diversity in others

Cohousing Questionnaire

Wonderland Hill has a questionnaire for potential buyers of a cohousing community that asks them to make their own decision about whether the cohousing lifestyle is for them. The following list paraphrases some of the attitudes inherent in cohousing for people to ponder to determine whether cohousing is for them:

- I respect other spiritual paths and do not hold mine as the only one.
- I value a sense of community with others.
- I appreciate diversity in others and in the community.
- I am interested in trying new experiences.
- I want to continue learning.
- I am open to change in the community and in myself.
- I am willing to participate in community activities.

- I value the environment and take action to help.
- I want to help to create the community I will live in.
- I have or would like to have a regular spiritual practice.
- I would like to volunteer to help others.
- I try to be as physically active as possible.
- I am willing to explore the mysteries of aging and death with others.
- I am willing to develop my talents.

As author Kephart toured Wild Sage, he met Eva, one of the older residents. They stood outside the common house looking, while they talked, at the mountain view to the southwest, unobstructed by the new Silver Sage buildings being framed. Eva told him of her dedication to remain in Wild Sage, rather than joining the "old folks" she envisioned moving into that community across the street.

Builders should be aware of these sentiments in homebuyers. Many of them are highly sensitive to being considered "old," especially baby boomers, and avoid anything that would label them in that fashion. Being part of an intergenerational community is important to these people. On the other hand, just because a community is designed for 50+ homebuyers, it doesn't need to project an image to match the most negative stereotyping of people who happen to be over 50.

One group of 50+ people trucked their kayaks to canyon country to enjoy the outdoors, the beautiful river canyons and to shoot the rapids. They are over 50, but they are athletic, healthy, and enjoying life like never before. Pictures of such groups of active people enjoying physically challenging activities are great images for builders of cohousing, or any kind of community for people over 50, to display and use in promotional and sales materials, versus allowing the uninformed to imagine the worst and then spread that image through word of mouth.

ElderSpirit, Abingdon, Virginia

ElderSpirit is one of the first age-qualified cohousing communities. It is being constructed by a nonprofit, tax-exempt corporation. Phase I of 25 homes was completed in June 2006; 12 homes are for rent and 13 for purchase. Phase II is the common house, which has four one-bedroom apartments for rent to low-income people.

The construction is simple, economical, and traditional in style, as opposed to the contemporary styling of Silver Sage. The variety of styles and price ranges of cohousing communities is one of the great things about them; every cohousing development has its own distinct style of architecture and living. The values upon which ElderSpirit was formed are clearly stated:

- Spirituality
- Mutual support

- Simple lifestyle and respect for the earth
- Arts and recreation
- Health
- Care during illness and dying
- Mutual assistance

For more information go to the ElderSpirit Web site: www.elderspirit.net. Mary Dene Peterson, founder of ElderSpirit, may respond personally. She has been very open and helpful to the authors in their research for this book and in sharing her information and photographs.

For additional information on cohousing there is an abundance of sources on the Internet. You can start with the following:

- The Cohousing Association of the United States: www.cohousing.org
- McCamant and Durrett Architects/The Cohousing Company: www.cohousingco.com
- Abraham Paiss and Associates: www.abrahampaiss.com

B

LifeCenter Plus

David Hall, founder of LifeCenter Plus in Hudson, Ohio, has created a model fitness and health delivery system he believes can be replicated in 50+ communities. LifeCenter Plus in Hudson has 187 staff, 6,500 full members, and serves more than 1,200 people per day. It is a breeder facility that will offer training and consultation to builders and developers of 50+ communities. The following is his description of the concept for improving the life and health of persons 50+, which he wrote especially for this book.

LifeCenter Plus is a 103,000-square-foot unique health and fitness center ranked in the top 100 in North America and located in Hudson, Ohio. LifeCenter Plus is a family-focused center with an older adult emphasis. A diverse array of programs is offered in 180 classes each week by a staff of 187 people, serving an average of 1,200 members per day, with one-third of those served being 70+ years old. David Hall is also the owner of real brokerage and development company with 30+ years of experience in site selection. This unique blend of backgrounds coupled with post-graduate studies in gerontology adds dimension and unique perspectives to the 55+ market adaptations.

The Amenity Package: From Bricks and Sticks to Lifestyle

(David Hall, LifeCenter Plus)

Although understanding the market need and the individual perspective are not rocket science it is also not necessarily obvious. The aging of America, or, if you like, the gray-ing of America, is happening all around us with first-time-ever demographic events throughout. But the time frames, successful aging perspectives, and the necessary net-work to provide essential services seem to get lost or go unacknowledged in the housing development attempts of today. Consider this: you will likely spend one-third of your life past the age of 60! And during this latter one-third of your life you will likely experience some of life's biggest social changes and psychological and physical challenges:

death of a child, death of a spouse, moving, loss of a job, loss of physical capabilities, and more.

As Art Linkletter once said, "Retirement isn't for sissies!" You need a thorough plan and ways to implement it. Here lie the real challenges and opportunities for the latter third of your life. Here also lie the biggest opportunities for the developers of older adult housing. Let's look at the needs, the nuances, and the opportunities.

Everywhere you turn, some financial planning service or insurance company is reiterating the importance of financial planning, insurance products, wills, and trusts. But what about a plan for the quality of life pieces? What, for instance, will you do for the following:

1. *Intellectual stimulation:* Basket weaving and Bingo didn't win your heart at age 14; they sure won't win your heart 60 years later!
2. *Socialization:* Who can live successfully without other people?
3. *Affirmation:* What are the patterns that affirm you and how will you replicate them in retirement?
4. *Psychological reward:* What rewards satisfy your drives and how will you get them?
5. *Physiological enhancement:* Lose your physical mobility and you lose your independence, and spend a lot of money prematurely or even unnecessarily. What will you do?
6. *Spiritual/emotional balancing*: This is an essential component to maintaining health while fortifying the immune system. How will you incorporate this critical component into your personal plan? How will you achieve meaningful involvement without it?

What are your personal options, your alternatives, and the best sources for your combined personal plan? How do you get access to the necessary services and, even more challenging, at affordable prices? Here's where it gets even more dependent on economies of scale and networking. And here's where it gets extremely difficult to accomplish as an individual. Without an organization to assemble the pieces, network them, and maintain them for easy access by the individual, the challenge may not be met. Like in any business model, spreading the costs to make it affordable is essential and networking is often the answer. The developers of housing for the 50+ market may be one of the few parties skilled and motivated to help the individual obtain these essential services. The completeness with which the developer provides for the above "bundle" of needs is crucial, and directly relates to both the success and the market absorption rate for the development.

Successful real estate developers for years have evidenced the right networking approach wherein they brought together many disciplines (legal, site analysis, geotechnical, wetlands, engineering, banking, real estate brokerage, construction, etc.) to create the ultimate product. So, too, for the successful provision of amenities and programming, think networking! Without the amenities and programming, successful adaptation to the older adult housing opportunities is really difficult. After all, homebuyers in this age range typically already have a package of bricks and sticks.

Your package may be slightly or even vastly different, but will the bricks and sticks package carry the day? There certainly exist some obvious design adaptations, like levered door handles, higher lighting levels, wider hallways and doors, raised electrical outlets, and single-level floor plans, but the real differentiation exists in the program and amenity package.

Let's take a look at each of these components and highlight a few of the ways by which they may be best accommodated in order to illustrate the process and approach which is being advocated.

1. Intellectual Stimulation

Working or retired, bored is bored, and boredom is not an enticement for the rest of your life, or even tomorrow. Intellectual stimulation is as diverse as the mix of people attempting to find it, and so too must be the sources. Rather than attempt to network every possible source, think networking the systems and the suppliers. Systems, like Internet access, are obviously key and basic. In a more personal and local way, existing community colleges and universities, clubs, organizations, and social clubs formed around interests can be helpful as well. Fortunately for the programmers, most people who desire and pursue these needs are typically both intellectually capable and likely to be already involved to a great extent. The disruption to some of these existing networks typically comes when there is a significant change/precipitating event in the person's life (death of a spouse, moving, change in health). Since these precipitating events often coincide with buying moments and changes of residences, a concerned and skilled concierge can be invaluable. This concierge, when properly trained, can assist on an individual basis to help the resident reassess and reconnect their intellectual desires with the networks at hand.

2. Socialization

That no man is an island is obvious. Getting man off the island and socially reconnected as he transitions though the "latter third" is the challenge. Spousal loss alone brings a need for new connections and "bridges." And even if that need isn't for another spouse or companion, there is still a need for replacement or substitution of that social function. During and shortly following the funeral, a lot of support is offered and then wanes with the passage of time. Slowly a psychological "threshold" emerges, where the resident becomes aware of a new need and then seeks a new answer. What a great time for the personal concierge, for new introductions, personal cheerleading, and confidence boosting or maybe even a new environment.

3. Affirmation

This component is one of the biggest, most important, and most difficult to understand. Let's start from a very personal point of view, your point of view. What is it that affirms you? What makes you feel important, worthwhile, and gives you meaningful involvement? What rewards you psychologically and makes you feel like you've done your best to meet your goals and desires? When you have uncovered the answers to all of those amalgamated social and psychological, lifetime-developed stimuli, then you will both understand affirmation and its importance in everyone's life. More importantly, this process of affirming yourself, or participating in functions and events that affirm you, may well be the most important component for your personal planning.

Consider all of the people whom you regard as living a "successful" retirement and then ask yourself, how many of these people are replicating the same social and psychological patterns which earlier affirmed them during their "working" career? Is the "care-giver" doctor or nurse still taking care of others, maybe in a different venue, but still care giving? Is the teacher still "educating"? I know we all assume the salesman will always be selling.

Uncovering, or helping the residents to uncover the patterns and successfully replicating those patterns in their new phase of life is the art of the concierge, and a powerful and meaningful enhancement to a lifestyle. Connecting them to a new network, where they can be meaningfully involved and replicate their personal patterns which affirm them, is the ultimate in amenity service and packaging. On your way to helping them be affirmed and fulfilled in a meaningful way, don't get lost in the "rocking-chair reward" myth; it's only an occasional break, not a lifestyle.

4. Psychological Reward

Stimulus, response, and reward are the components, and *Psychology 101* tells the story. The laboratory rat, driven by hunger (the stimulus), presses a metal bar (the response), and receives the kernel of corn (the reward). If, after every pressing of the bar, the kernel of corn falls, it is called continuous reinforcement. If only on occasion a kernel falls and falls with no predictable pattern, then it is called intermittent reinforcement, the hardest behavior to extinguish and the force that brings us back to the slot machines in Vegas with more frequency than is financially appropriate.

When we live in a work-focused society which rewards involvement and contribution, and misunderstands noninvolvement or zero contribution as slothful and lazy, it compromises retirement and confuses the participants.

After 65 years of involvement in this type of society and the bulk of those years involved in some occupation, reward oftentimes becomes quantified and the only tangible evidence of involvement. Success becomes defined by occupation and the resulting measurable achievements, that is, products manufactured, policies sold, and customers served. Dignity then becomes a function of occupation and when the occupation ceases, the dignity gets disconnected and the challenges emerge. The husband comes home to retirement, does the "honey do" list, OD's on fishing and golf, then starts to tell the wife how to run the house! She's thinking, "I married this guy for life, but not for lunch!" and the friction starts. The stress of losing his occupation, his dignity, his meaningful involvement, and some way to measure his involvement, is all too much to bear. The stress of the entirety of his situation breaks down his immune system and disease sets in to diminish his quality and shorten his life.

Finding and measuring alternative involvements is essential to continuing the reward patterns and prolonging his life and his wife's sanity. And remember, it's not what you measure, but the measurement process that is critical.

The absence of measurement and psychological reward breaks down the immune system, which often leads to unnecessary disease. The psychological study, "Ulcers in Executive Monkeys," showed what happens when monkeys, conditioned to and rewarded by tasks, are deprived from psychologically reinforced involvement. The study provided insights as to why ulcers are developed on weekends, vacations, and retirement!

Most importantly, when planning a 50+ community, research those ways which are available to provide meaningful and measurable ways to reward resident involvement in an intermittent manner. If they don't exist, then program measurements into your walking path, your fitness center, your activity involvements. First measure, then outwardly reward, those involvements which are necessary. This continued reinforcement is essential for the health of the residents and the project.

5. Physiological Enhancement

Lose your physical mobility and lose your independence. It's not about aging, it's about disuse. Don't fatalistically accept your lot in life based on age. We are now just learning what it means to be age 80 and beyond. The human body responds favorably to physical stress and if you haven't seen first-hand the results of aging and exercise, go visit your local International Health Racquet and Sportsclub Association (www.ihrsa.org) club.

Fitness facilities obviously correlate to the physiological enhancement needs and, if done properly, they go a long way to providing socialization and a venue for the individual to measure results and psychologically reinforce themselves for their involvement. The challenge for an effective fitness component rests in the economy of scale and pricing challenge. Today's successful health and fitness facilities typically have 60,000 square feet or more and require thousands of paying members to accommodate affordable monthly dues. Smaller, more focused facilities are possible, but with narrowly defined and higher prices for service components and an emphasis on personal training and physical therapy. This economy of scale narrows the site selection process and more strongly correlates the project size with the fitness facilities, whether existing or to be built. The effect for smaller housing projects is that they must locate near existing and accessible facilities, while providing some of the common area space for more flexible space needs. Open rooms, where a variety of programs can be offered, are central to the financial and programming success of any center. To provide a wide variety of uses is in great part a function of having qualified people with strong programming skills, who coordinate the diversity of programming necessary to have broad market appeal. To successfully meet the health and fitness needs of the residents, an array of programs, including such topics as nutrition, new medical advancements, meal planning, stress relief, sleep techniques, and others, are needed and generally provided by area providers.

Where economies of scale do exist, the health and fitness facilities should be located on the perimeter of the development to be, and be perceived as, accessible to the surrounding community. A big part of success in today's fitness facilities is to be more inclusive rather than exclusive. Intergenerational programming appeals to the older adult market because the grandchildren can mix with the grandparents and all ages in-between.

Projects that restrict children may well appeal to a select group of homebuyers; however, the common amenities, with their operating economies of scale, are more successful when all day, year-round programming is provided for people of all ages.

The physical results of the combined programming should address cardiovascular or aerobic exercise, together with strength training, balance, and flexibility. When these physical components are combined with the appropriate nutritional and medical components, powerful results and impressive competitive advantages can be achieved.

6. Spiritual/Emotional Balancing

In their book, *Younger Next Year*, authors Henry S. Lodge, M.D., and Chris Crowley detail the importance of exercise, diet, and aging to achieve good health and an effective maintenance of the immune system. With great detail and impressive medical research, they stress the importance of combining exercise, diet, and emotional and spiritual elements to live beyond 80 looking and feeling like 50. It's cutting-edge research presented in a motivating way and with great timing. It's a worthwhile read and a road map for successful aging, and a part of the retirement planning process so necessary for quality and longevity. Without the spiritual balance, the human organisms don't maintain a harmony and balance so essential to good health and the preservation of good health.

Many organized religious groups supply most of the components that are essential for this element, and this author does not intend to interfere or disrupt those processes. The importance of this paragraph is simply to emphasize the importance of this element in both the overall planning and implementation of a 50+ community and to emphasize the importance of inclusion of this component in the network. An effective network to supply the target market demographic profile to be served will certainly include this component to assure success.

Summary

Understanding the diverse individual needs of the aging American marketplace is essential to designing and providing successful projects with the most appeal, the best absorption rates and the highest profitability. The array and complexity can most profitably be provided by networking with established community providers. Operational economies of scale to create and maintain critical services must be adhered to quite closely, given the importance and cost of these critical services. Absent the integrative and networked approach as advocated to meet the needs outlined, the costs of health care and other care will result in an ever-increasing cost structure, which will ultimately and severely limit the older adult marketplace.

City of Portland, Oregon, Title 33, Chapter 33.205: Accessory Dwelling Units

333.205.010 Purpose
Accessory dwelling units are allowed in certain situations to:

- Create new housing units while respecting the look and scale of single-dwelling development;

- Increase the housing stock of existing neighborhoods in a manner that is less intense than alternatives;

- Allow more efficient use of existing housing stock and infrastructure;

- Provide a mix of housing that responds to changing family needs and smaller households;

- Provide a means for residents, particularly seniors, single parents, and families with grown children, to remain in their homes and neighborhoods, and obtain extra income, security, companionship and services; and

- Provide a broader range of accessible and more affordable housing.

33.205.020 Where These Regulations Apply
An accessory dwelling unit may be added to a house, attached house, or manufactured home in an R zone, except for attached houses in the R20 through R5 zones that were built using the regulations of 33.110.240.E, Duplexes and Attached Houses on Corners.

33.205.030 Design Standards
C. Requirements for all accessory dwelling units. All accessory dwelling units must meet the following:

1. Creation. An accessory dwelling unit may only be created through the following methods:

   a. Converting existing living area, attic, basement or garage;

b. Adding floor area;

c. Constructing a detached accessory dwelling unit on a site with an existing house, attached house, or manufactured home; or

d. Constructing a new house, attached house, or manufactured home with an internal or detached accessory dwelling unit.

2. Number of residents. The total number of individuals that reside in both units may not exceed the number that is allowed for a household . . .

3. Parking. No additional parking is required for the accessory dwelling unit. Existing required parking for the house, attached house, or manufactured home must be maintained or replaced on-site.

4. Maximum size. The size of the accessory dwelling unit may be no more than 33% of the living area of the house, attached house, or manufactured home or 800 square feet, whichever is less . . .

D. Additional requirements for detached accessory dwelling units. Detached accessory dwelling units must meet the following.

1. Setbacks. The accessory dwelling unit must be at least:

a. 60 feet from the front lot line; or

b. 6 feet behind the house, attached house, or manufactured home.

2. Height. The maximum height allowed for a detached accessory dwelling unit is 18 feet.

3. Bulk limitation. The building coverage for the detached accessory dwelling unit may not be larger than the building coverage of the house, attached house, or manufactured home. The combined building coverage of all detached accessory structures may not exceed 15 percent of the total area of the site.

33.205.040 Density

In the single-dwelling zones, accessory dwelling units are not included in the minimum or maximum density calculations for a site. In all other zones, accessory dwelling units are included in the minimum density calculations, but are not included in the maximum density calculations.

Glossary

"To get the right word in the right place is a rare achievement."
—Mark Twain

AARP—Formerly American Association of Retired Persons, now goes by just AARP. Nonprofit organization of some 40 million members, committed to making life better for people age 50 and over. Known for its powerful lobbying and extensive education services. www.aarp.org.

Accessory apartment—Another term for an accessory dwelling unit (ADU) or detached accessory dwelling unit (DADU) used in some city zoning ordinances, for example Austin, Texas.

Active adult—Refers to communities that are age-restricted, usually offering features, amenities, and socializing opportunities for residents age 55+.

ADU—Accessory dwelling unit, a small cottage separate from the house but on the same property which can house a parent or two, an adult child, or a visitor or two close by without interfering with the occupants of the main house. (See also DADU)

Age-qualified—Same as age-restricted.

Age-restricted—A community in which at least one person who owns or rents each housing unit must be a certain age or older, usually age 55 or older. Usually applies to some percentage of residents, frequently 80 percent, sometimes 100 percent. Same as age-qualified.

Age-targeted—A community where housing units are not age-restricted but are targeted to appeal to people age 55 and over by the design, features, and amenities.

Aging in place—The term that refers to the ability for persons of any age to live in a home or for the same persons to be able to live in a home from when they are young until they are old.

AOA—Administration on Aging. www.aoa.gov.

ASA—American Society on Aging, an advocate group.

Baby boomer—The term in common usage for the generation born between 1946 and 1964.

Boomer—Baby boomer.

Carriage house—A small building usually near a large residence or part of an estate, used for keeping coaches, carriages, or other vehicles; also called a coach house.

CBO—Congressional Budget Office. Assists Congress with objective, nonpartisan information and analysis regarding the federal budget. www.cbo.gov.

CCRC—Continuing care retirement center.

CCTV—Closed-circuit TV.

Cohousing—A living arrangement that combines private living quarters with common dining and activity areas in a community whose residents share in tasks such as errands, childcare, and healthcare.

Comfort-height toilet—Toilet that is the approximate height of a chair, about 16 1/2 inches high, a few inches higher than what is generally installed.

Continuing care retirement center—Complex that offers health-related services as well as residences. May include some or all of the following: independent living, assisted living, Alzheimer's and dementia care, and skilled nursing care.

DADU—A detached accessory dwelling unit (see ADU).

EBRI—The Employee Benefit Research Institute. A nonprofit, non-partisan research group whose mission is to contribute to, encourage, and enhance the development of sound employee benefit programs and sound public policy through objective research and education.

Eisenhower Generation—(See silents) The generation born between 1926 and 1945.

Energy Star—An energy usage standard by the United States Environmental Protection Agency and the U.S. Dept. of Energy.

FHA—The Federal Fair Housing Act, also used for the Federal Housing Administration.

GI generation—The term used for the generation of persons born before 1925.

Gen X or Generation X—The generation following the baby boomers, especially Americans and Canadians born in the 1960s and 1970s.

GPS—Global positioning system, a global satellite system that aids navigation in real time.

Household—A residential unit and its occupants. Can be one person living there or multiple.

IBS—International Builders' Show, an annual convention and expo sponsored by the National Association of Home Builders.

Intergenerational—Not restricted to members of one generation or age group.

LEED—Leadership in Energy and Environmental Design. A program of the U.S. Green Building Council, rating green elements of building design and development.

Lifestyle community—A way of life or style of living that reflects the attitudes and values of a person or group.

LOHAS—Lifestyles of Health and Sustainability. A private consumer education group.

MetLife MMI—MetLife Mature Marketing Institute. www.maturemarketinstitute.com.

MFM—A house that has the master bedroom on the first floor with the other bedrooms upstairs.

NAHB—National Association of Home Builders. www.nahb.org.

OPEC—Organization of Petroleum Exporting Countries. A cartel of oil-producing countries. www.opec.org.

PCBC—Formerly Pacific Coast Builders Conference, now just PCBC. Group of building-related professionals and related persons and organizations committed to "advancing the art plus science of community building." www.pcbc.com.

POD—A landplaning term for a section of a larger master plan, one that is not detailed with roads and lots.

Pogo—Popular central character in Walt Kelly's Pogo cartoon, which ran in daily newspapers from 1948 to 1975. Known for his satire and wisdom relating to social and political issues.

Quilting—A landplaning method of mixing lot sizes and building types throughout a new community plan, usually done to incorporate diverse sizes and prices of new homes within a community.

Ranch—A common term used for a single-story house, also known as a "rambler." Can also be a "flat" when one residence in a larger building contains multiple residences.

Senior—Thought to be a kinder, gentler word than elderly or old person. A word you never want to use when referring to a baby boomer.

Silents—(See Eisenhower Generation) The term used for the generation born between 1926 and 1945.

Social networking—Refers to going to Web sites where people organize themselves into groups for the purpose of interacting. People may be members of many groups within social networking sites. Examples are Facebook, LinkedIn, and Twitter. New ones pop up frequently.

Task lighting—Lighting designed to illuminate the area where people do specific tasks. It enhances general lighting, which alone often shines from behind the person, creating shadows over the work area.

TND—Traditional Neighborhood Development, meaning a development that mirrors past community designs in form and style.

TOD—Transportation Oriented Developments. Residences are within walking distance of mass transit hubs or stations as well as restaurants, services, and shopping.

Universal design—A form of design the purpose of which is to allow persons of all ages with any of a multitude of disabilities and health conditions to live and move easily and comfortably without barriers.

USGBC—United States Green Building Council. The creators and managers of the LEED rating system for green building.

Index

A

AARP, 198, 205, 241, 269

Abingdon, Virginia, 60, 223, 259–260

Abraham Paiss and Associates, 260

Accessibility:
  in bathrooms, 195
  of residential buildings, 80
  and universal design, 75–77

Accessory apartments, 269

Accessory dwelling units (ADUs), 17
  in future of homebuilding, 204–205
  market niches for, 226
  offices in, 197
  for three-generation families, 124–127
  zoning of, 128–130, 267–268

Active adult communities, 11
  condos and apartments in, 142, 143
  healthcare at, 207, 208
  market research for, 215–216
  for travelers, 185
  two-story homes in, 122–123

Active adults, 269

Activity centers (kitchens), 83

Activity level, of boomers, 241, 259

Adaptable design, 79

Adjustable shelves, 86

Administration on Aging (AOA), 270

ADUs (see Accessory dwelling units)

Affirmation, 262–264

Affordability, clustering and, 172–174

Age Power (Ken Dychtwald), 133

Age Wave, 133

Age-qualified housing, 25

Age-restricted housing:
  in Columbia, Maryland, 163–164
  communities of, 25, 32, 35
  condos and apartments as, 147–148
  market niches for, 224
  zoning rules for, 23

Age-targeted housing:
  communities of, 24–25, 30, 35
  condos and apartments as, 147–148

Aging in place, 63, 74–75, 251

Aging in Place Guide (Louis Tenenbaum), 81

Air heat exchangers, 70

Air infiltration, 68

Air quality, 70, 71

Ambassador programs, 240

Ambient lighting, 91

Amenities:
  of condos and apartments, 152–155
  in the design process, 175–181
  improving use of, 137
  invisible, 115
  at LifeCenter Plus, 261–266
  of lifestyle communities, 28–31
  market research about, 170
  of multigenerational communities, 107, 110
  of university-affiliated communities, 53

American Society on Aging (ASA), 270

Americans with Disabilities Act of 1990, 75, 80

Americans with Disabilities Act Amendments Act of 2008, 75

Anthem Highlands (community), 32

Anthem Parkside (community), 32, 107, 218

Anthem Ranch (community), 32, 218

AOA (Administration on Aging), 270

Apartments (*see* Condos and apartments)

Appliances, 81, 84–85, 190

Arbors at Bridgewater Crossing (community), 30–31, 173, 177–178, 198

Architects, preconceptions of, 113–116

Architecture:
  for active adult apartments, 152
  and the design process, 181–183
  for market niches, 229–230
  and solar orientation, 66

Artisan Homes, 148, 217, 242

The Artisan Lofts on Osborn, 41, 159, 217, 242

ASA (American Society on Aging), 270

Assisted living facilities, 208–209

Athletes, retirement communities for, 221

Atlanta, Georgia, 40, 42

Attics, 137, 198

Attitudes, of boomers, 9–10

Auburndale, Massachusetts, 50, 52

Aurora, Colorado, 164

Avatar (company), 28

Ave Maria, Florida, 32, 33, 223, 224

B

Baby Boomer Report, 26, 184, 196, 210, 217–218
Baby boomers (*see* Boomers)
Badler, Gerard, 51
Bainbridge Homes, 37
Balconies, 155
Balfour Senior Living (community), 53–54, 149
Bally's Fitness, 180
BankAmericard, 10
Baruch, Bernard M., 1
Basements, 137, 197
Batavia, Illinois, 174
Bathrooms:
    design options for, 194–196
    lighting for, 91
    preferred number of, 146–147
    technology in, 102
    universal design for, 87–90
Bathtubs, 87, 89, 195
Bedrooms:
    in active adult housing, 124
    design options for, 193–194
    layout of, 8
    location of, 199–200
    preferred number of, 146–147
    in resort-styled communities, 110
    in second homes, 120
    separate, 198
    technology in, 101–102
    in townhouses, 122
    in traditional neighborhood developments, 112
Behaviors, of boomers, 9–10, 19–21, 234–235
Belmar (community), 43, 151, 161–164
Berkus, Barry, 162
Big box home design, 134–135
Bike paths, 179
Binder, John, 68
The Black Swan (Nassim Nicholas Taleb), 202, 209
Blake, Deborah, 240
Blogs, 243–244
Blower door tests, 68, 69, 71–72

Bonus rooms, 134, 159
Boomers, 1–11, 233–245
    attitudes and behaviors of, 9–10
    comfort with technology, 95–97
    as consumers, 3–5
    demographics of, 2, 3
    designing and building for, 1–2
    diversity of, 210
    Eisenhower generation vs., 6–8
    generational roots of, 234–235
    identity of, 233–234
    lifestyle of, 10–11, 210, 242
    marketing for, 235–245
    mindset of, 235–236
    needs of, 8–9
    reported wealth of, 15–16
    returning to work by, 15, 183–184
    values of, 133–134
"Boomers and Technology" study, 95
Boomers on the Horizon (Margaret Wylde), 46, 187, 197, 230
Boomertising, 4, 8
"Boomies," 233
Boulder, Colorado:
    cohousing in, 57–60, 223, 226, 249, 254–255, 257
    university-affiliated retirement community in, 53
Bowen, Bryan, 59, 255, 257
Brecht, Susan, 217
Bright Move, 147
Britain, 233
Broomfield, Colorado, 32
Brunswick Hills, Ohio, 30, 173, 177–178, 198
Builders:
    advantages of green building for, 65
    knowledge about universal design of, 75
    market research for, 215–216
    preconceptions of, 113–116
    trend identification by, 201–202
    use of technology by, 96–98

Building:
    for boomers, 1–2
    modular, 135, 209
    (*See also* Future of building)
Building envelope, 68
Built Green Colorado, 74, 204
"The bulge," 233
Bunkrooms, 137
Business centers, 154–155, 184
Buyer preference studies, 168–169
Buyers:
    design options as viewed by, 198–200
    estimating number of, 217–218
    feelings of, 182–183
    information for, 238–240
    input from, for cohousing, 59
    niche markets of (*see* Market niches)
    pain points of, 138
    personal expression of, 133
    satisfaction surveys for, 236
    size preferences of, 230
    technology as viewed by, 97–98
    universal design as viewed by, 75–77

C

Cabinets:
    in bathrooms, 89–90
    in kitchens, 85–87
    in universal design, 81, 83
Cambridge Homes, 27
Campus Continuum LLC, 51
Canada, 233
Cappadocia, 93
Carbon monoxide monitors, 101
Carillon (community), 27
Carle, Andrew, 50, 55
Carlson, Rich, 240
Carriage houses, 121, 124, 129
    [*See also* Accessory dwelling units (ADUs)]
Casement windows, 82
Casitas, 196

Catholic Church, 32, 223
CBO (Congressional Budget Office), 270
CCCs (Citizen-Created Communities), 209
CCRCs (*see* Continuing care retirement communities)
Ceilings:
  of condos, 159
  of single-family homes, 130–131
Cell phones, 206, 234
Center for Universal Design, 77
Centex Homes, 176, 178
Change, forces of, 211–212
Chapman, Jim, 42, 153, 159–160, 176
Charettes, design, 236
Chicago, Illinois, 40, 148, 149, 221
Children:
  and age-restricted housing, 23, 25
  and cohousing, 250
  in multigenerational housing, 32, 34, 210
  as visitors, 31, 137
Cincinnati, Ohio, 179–180
Citizen-Created Communities (CCCs), 209
The City of Legends, 221
Civil Rights Act of 1964, 9
Classic Residence by Hyatt, 50
Classrooms (in community spaces), 184
Clermont, Florida, 221
Cleveland, Ohio, 40
  downtown population of, 149
  suburban city centers near, 163
  volume of development in, 133
Clinger, David, 118
Club 50 Fitness, 180
Club rooms, 152–154
Clubhouses:
  for apartments and condos, 152–154
  in design process, 175–178

in resort-styled communities, 110
Clustering, 172–173
CNU (Congress for the New Urbanism), 44
Co-care, 58, 251–252
Cohousing, 210, 247–260
  and co-care, 251–252
  and communes, 4, 5
  creating, 256–257
  at ElderSpirit, 259–260
  guidelines and characteristics of, 248–249
  market niches for, 223–225
  at Munksogard, 252–254
  and needs of seniors, 249–250
  neighborhoods for, 55–60
  reasons for choosing, 250–251
  senior, 250, 251
  at Silver Sage, 254–255
  single-family homes for, 112
  Web resources on, 260
  by Wonderland Hill Development Company, 257–259
Cohousing (Kathryn McCamant and Charles Durrett), 55, 247
Cohousing Association of the United States, 260
Cold War, 9
Colleges (*see* University-affiliated retirement communities)
Colorado Land and Building Company, 38–39
Colorado Land and Home Company, 171
Colorado State University, 184
Columbia, Maryland, 153, 163–164
Comfort-height toilets, 270
Commercial entities (in town centers), 45
Commercials, television, 245
Common houses, 56, 112, 248, 255
Communes, 4–5

Communication:
  of marketing messages, 236, 242–245
  technology for, 98, 99, 101
Communities:
  developments as, 227
  resort-styled, 110–111
  (*See also* Neighborhoods)
Community centers, 177–178
  (*See also* Clubhouses)
Community identity, 227–228
Community vision, 172, 228–229
Compact fluorescent light bulbs, 70
Compact living (in traditional neighborhood developments), 44, 45
Competitors, market research about, 236
Computer rooms, 155
Computers, boomers' use of, 98
Concierges, 153, 263
Condos and apartments, 141–165
  age-restricted/age-targeted/ lifestyle-targeted, 147–148
  amenities of, 152–155
  design considerations for, 150–160
  design principles for, 159–160
  elements of, 155–158
  and home ownership, 146–147
  location of, 148–151
  market trends for, 141–143
  parking for, 158–159
  as second homes, 120–121
  similarities and differences between, 144–146
  storage in, 158
  in suburban city centers, 161–165
  urban housing models for, 160–161
Congress for the New Urbanism (CNU), 44
Congressional Budget Office (CBO), 270

Consumer confidence, recession and, 13
Consumers, boomers as, 3–5
Continuing care retirement communities (CCRCs), 52–53, 207
Continuum Partners, 43, 162
Control, technology for, 98–99
Convenience, 3, 250
Cooktops, 84
Cooling systems, 64, 73, 181
Corners, building, 134, 135
Corte Bella (community), 30, 223
Cosmopolitan Club, 149–150
Costs:
  of active adult homes, 131, 132
  of cohousing, 250
  of elevators in homes, 132
  life cycle, 68
  of one-story vs. two-story homes, 123
  of two-story single-family homes, 131, 132
Cotati, California, 57, 249
Countertops:
  in bathrooms, 88
  in kitchens, 85, 191, 192
Courtyards, 53, 106, 131, 199, 200
Courtyards at Rolling Hills (community), 38
Covell, Chuck, 238–240
Credit cards, 10
Crest Apartments, 161
Cross, Tracy, 27
Crowley, Chris, 266
Cuban Missile Crisis, 9
Cul-de-sacs, solar orientation for, 66
Cultural niches, 222–224
Cyclic life paradigm, 7

D
DADU (see Detached accessory dwelling unit)
Dallas, Texas, 228

Darwin, Charles, 13
Davis, California, 60
Del Webb, 27, 30, 32, 107, 159, 160, 171, 198, 217, 218, 223
Demographics:
  of boomers, 2, 3
  in market research, 236
  of renters, 142
Denmark, cohousing in, 250, 252–254
Densities, housing, 36–38, 123
Denver, Colorado:
  accessory dwelling units in, 129, 130, 205
  apartment developments in, 144–146, 149–150, 161
  cohousing in, 59
  downtown population of, 40–41, 148–149
  infill neighborhoods in, 35–38
  transit-oriented design in, 47–49
Denver Council of Regional Governments (DRCOG), 205
Deprivation, 235
Design charettes, 236
Design inefficiencies, 115
Design options, 187–200
  for bathrooms, 194–196
  for bedrooms, 193–194
  buyers' views of, 198–200
  for condos and apartments, 150–160
  for great rooms, 187–192
  for neighborhoods, 23–24
  for offices, 196–197
  for storage, 197–198
Design process, 167–185
  architecture for homes in, 181–183
  for boomers, 1–2
  for cohousing developments, 56–57
  community amenities in, 175–181
  community vision in, 172

homes and community synergy in, 183–185
  for market niches, 227–230
  market research in, 167–170
  market studies in, 170–172
  neighborhood considerations in, 172–174
  for traditional neighborhood developments, 44
Detached accessory dwelling unit (DADU), 270
Development(s):
  as communities, 227
  political environment for, 38–39
Dickerson, Cynthia, 138
Dimmers, 70, 91, 102
Dining rooms, 102, 120, 188–191
Disabilities, 75, 77
Disease, psychological reward and, 265
Dishwashers, 85
Disney World, 3
Disneyland, 3
Diversity (of boomers), 210
DiVosta Homes, 223
Doherty, Paul, 96–98
Domino's Pizza, 32, 223
Door handles, 86
Doors and doorways:
  of bathrooms, 87, 195
  glass, 203
  universal design for, 81–82
Downsizing, 138
Downtowns (see Urban areas)
Drawer handles, 86
DRCOG (Denver Council of Regional Governments), 205
Drip irrigation, 66, 73
DTJ Community Architecture and Planning, 53
Dual-flush toilets, 73
Ductwork, 68–70
Duplexes, 121
Durrett, Charles, 55, 56, 59, 225
Dychtwald, Ken, 7, 133

E
Eames, Charles, 63
Earth Day, 64, 92
Easements, 119
Eating spaces (in kitchens), 86
EBRI (Employee Benefit
    Research Institute), 270
Eco-Village (community),
    42, 160
Education, 51, 226, 250
Ehlers, Janis, 239
Eisenhower generation, 6–8,
    18, 234
Elburn, Illinois, 47–48
ElderSpirit (community), 60, 223,
    259–260
Electrical outlets and panels,
    locations of, 80
Elevators:
    in accessory dwelling
        units, 127
    in condos and apartments, 159
    in infill housing, 36–39
    in multigenerational
        housing, 34
    in single-family homes, 49,
        124, 126, 131–133
    in townhouses, 121, 122
    in traditional neighborhood
        developments, 45
    in universal design, 81
eMarketer, 238
e-marketing, 239
Emotional balance, 262, 266
Employee Benefit Research
    Institute (EBRI), 270
Emrath, Paul, 138
Energy conservation:
    and cohousing, 250
    floor plans for, 67
    in green building design, 64,
        68–72
    savings from, 65, 93, 204
Energy Star program, 204
Engagement, in marketing,
    239–240
Engle Homes, 64

Entertainment, technology for,
    99–102
Entitlement process, 229
Entries:
    technology for, 102
    in traditional neighborhood
        developments, 111
    universal design for, 81–82
    zero-step (no-step), 76, 80, 81,
        93, 110
Envelope, building, 68
Environment, technology for
    control of, 99
Equitable use (in universal
    design), 77
Equity (in condos), 144
Erie, Ohio, 37
Esprit Homes, 93, 151
Event marketing, 240
The Evergreens, 153
Exclusivity, 224
Eyesight, 90

F
Facebook, 24, 244
Fair Housing Act of 1968, 25, 80
Families:
    nuclear, 249
    size of, 218
    three-generation, 125–127, 203
Family rooms, 139
Fannie Mae, 141
Fans, on Facebook, 24, 244
Feasibility phase (for
    cohousing), 256
Feasibility studies, 235–236
Federal Fair Housing Law (FHA),
    221, 222, 270
Federation of Communities in
    Service (FOCIS), 60
Feelings, of buyers, 182–183
FHA (see Federal Fair Housing
    Law)
Finish materials, 181
Fitness centers:
    for apartments and condos,
        154, 160

in clubhouses, 177
in design process, 180–181
at LifeCenter Plus, 261–266
in multigenerational
    housing, 34
in traditional neighborhood
    developments, 45
in urban centers, 41
Fitness directors, 31
Five-star hotel concept (for
    condos), 160
Flexibility of use (in universal
    design), 78
Flexible space, 155–156, 193
Flooding, 93
Floor plans:
    of accessory dwelling units,
        128
    bedrooms in, 194
    in cohousing communities, 57
    of condos and apartments, 155
    courtyards in, 106
    dining and living rooms in, 189
    for energy conservation, 67
    great rooms in, 188
    and home size, 108–109
    in resort-styled communities,
        110, 111
    see-through, 27, 28, 198–199
    of townhouses, 122
    in traditional neighborhood
        developments, 112
    of two-story homes, 125, 126
    for universal design, 80
    in urban downtown com-
        munities, 41
Flooring, nonskid, 88
Fluorescent lighting, 70, 91, 92
Focalyst Insight Report, 92
FOCIS (Federation of
    Communities in Service), 60
Focus groups, 170–172, 236
Followers, on Twitter, 24, 244–245
Fort Collins, Colorado, 184
Fortenberry, Christine, 75
Frankfort, Illinois, 174
Freddie Mac, 141

Free generation, 8
Fresh air, tempering of, 70
Friends of Granny Flats, 130
Frog Song (community), 57, 249
Front-loaded garages, 122, 131
Fun zones, 162–163
Functional design elements, 181
Furnaces, 69, 70, 72
Future of building (see Trends)

G
Gainesville, Florida, 52
Galley kitchens, 190
Garages:
    for condos and apartments, 142
    front-loaded, 122, 131
    rear-loaded, 44, 46, 122
    and solar orientation, 67
    storage in, 197–198
    technology in, 102
    for townhouses, 122
    in traditional neighborhood
        developments, 44, 111
    and views from homes, 131
Gathering nodes, in cohousing,
    56–57, 248
Gay communities, 222, 225–226
General contractors, 160
General Growth Properties
    (GGP), 163
Generation X (Gen Xers), 18, 270
Generational roots (of boomers),
    234–235
The Geography of Nowhere
    (James Howard Kunstler), 5
George Mason University, 55
Georgetown 2 (community), 174
Geothermal heating and
    cooling, 73
GGP (General Growth
    Properties), 163
GI generation, 18
Glacier Circle (community),
    60, 254
Glare, minimizing, 91
Glass, low-E, 66
Glass doors, 203

Glendinning, Keith, 208
Glendinning, Ruth, 208, 209
Global positioning systems
    (GPSs), 206
Globalization, 211
Gold's Gym, 180
Golf cars (golf carts), 29, 180
Golf courses:
    in the design process, 175
    in lifestyle communities, 27, 29
    in market studies, 169
    and walking trails, 180
Gonzalez, Manny, 154, 158
Government (as force of
    change), 211
GPSs (global positioning
    systems), 206
Grab bars, 77, 81, 88, 90, 195
Granny flats, 17, 129, 130
    [See also Accessory dwelling
    units (ADUs)]
Grass Valley, California, 60
Gray gyms, 180
Great Depression, 6, 234
Great rooms:
    in active adult housing, 124
    in condos and apartments, 155
    design options for, 187–192
    in resort-styled communities,
        110
    in second homes, 120
    technology in, 100–101
Green, Kay, 190, 195
Green building design,
    63–74, 181
    for apartments and condos,
        157–158
    building orientation in, 65–67
    and energy conservation, 68–72
    in future of homebuilding, 204
    and green homebuilding, 73–74
    interest in, 92–94
    and mainstream green move-
        ment, 64–65
    and universal design, 63–64
    and university-affiliated retire-
        ment communities, 52

    and water conservation, 72
    at Wild Sage, 58
    by Wonderland Hills
        Development Company,
        257, 258
Green collar economy, 17
Green Home Guidelines, 74
Green movement, 4, 10, 64–65
Grey water collection, 73
Guest suites, 137
Guilt (as sales tool), 239
Gyms, 180 (See also Fitness
    centers)

H
Hall, David, 41–42, 180, 261
Hallways:
    lighting in, 91, 101
    universal design for, 80–82
Hampton, Lionel, xi
Handles, door and drawer, 86
Head, Heart, and Hands (Shari
    Leach), 258
Healthcare:
    and co-care, 252
    facilities for, 207–209
    monitoring of, 99
Healthful living, cohousing
    and, 251
Heating and cooling systems, 64,
    73, 181
Hidden Lake (community), 174
Highland Gardens (community),
    59
Highlands Group, P.C., 60
HOAs (home owners associa-
    tions), 152
Holy Cross College, 52
Holy Cross Village (community),
    52, 207
Home and community synergy
    (in the design process),
    183–185
Home Builders Association of
    Metro Denver, 74
Home Design Trends Survey, 203
Home offices, 184, 193, 196–197

Home owners associations (HOAs), 152
Home ownership, condos and, 146–147
Home sizes:
  and big box design, 134–135
  of condos and apartments, 150
  in future of homebuilding, 203
  minimums for, 229
  and rightsizing, 137–139
  trends in, 136–137, 230
Hong Kong, 48
Hord, Ed, 142, 147
Household, defined, 271
Housing densities, 36–38, 123
Housing for Older Persons Act of 1995, 25
"Housing for the 55+ Market" report, 20–21
Houston, Texas, 40, 149
Howard, Evelyn, 150, 153, 154, 158
HUD (see U.S. Department of Housing and Urban Development)
Hudson, Ohio, 41, 261–266
Huntley, Illinois, 27

I
IBS (International Builders' Show), 271
Identity:
  of boomers, 233–234
  community, 227–228
Immigration (as force of change), 211
Immune system, health of, 265, 266
Income (of market niche), 218, 219
"The Incredible Shrinking Boomer Economy" (David Welch), 19, 21
India, 191, 211
Indiana University, 52, 207
Inefficiencies, design, 115
Infill neighborhoods, 36–39

challenges with, 38–39
condos and apartments in, 151
as development opportunities, 36–38
market for, 149, 171
in resort areas, 118
in suburban city centers, 42
Information:
  marketing, 238–240
  perceptible, 78
Information phase (for cohousing), 256
In-law suites, 127 [See also Accessory dwelling units (ADUs)]
Installation process (for technology), 100–103
Insulation, 64, 68, 72
Intellectual stimulation, 262, 263
Interactivity, in marketing, 242
Intergenerational (term), 271
International Builders' Show (IBS), 271
Internet:
  and buying decision, 24
  marketing with, 236, 237, 242–245
Internet access, 97–98, 154
Invisible amenities, 115
Irrigation systems, rainwater, 66
Islands, kitchen, 83–84, 191, 192

J
Jayson, Sharon, 203
Jefferson, Thomas, 201
Jensen, David, 179
Johnson, Ric, 100

K
K. Hovnanian Homes, 148
Kaufmann, Michelle, 209
Kennedy, John F., 9
KEPHART, 36, 172
Ketchum, Idaho, 117, 118
Kicking Horse condominiums, 117

Kinesthetic experience (of homes), 182
King, Martin Luther, Jr., 9
Kissimmee, Florida, 180
Kitchen designers, 84
Kitchens:
  in cohousing communities, 112
  of condos and apartments, 158
  design options for, 189–193
  functional design elements of, 181
  islands in, 83–84, 191, 192
  lighting for, 91–92
  in resort-styled communities, 110
  in second homes, 120
  summer, 196–197
  technology for, 101
  in traditional neighborhood developments, 112
  universal design for, 82–87
Kriescher, Paul, 68
KTGY Group Inc., 154
Kunstler, James Howard, 5

L
Lady Lake, Florida, 29, 180
Lakewood, Colorado, 43, 151, 161–162, 164, 228
Land planning, 226–229
Lasell College, 50, 52
Lasell Village (community), 52
Late boomers (trailing-edge boomers), 1, 6
Laundry rooms, 158
Layouts (of kitchens), 83–84, 190
Leach, Jim, 58, 60, 257, 258
Leach, Shari, 60, 258
Leadership in Energy and Environmental Design (LEED) certification, 52, 204, 271
Leading-edge boomers, 1, 6
Leaks, energy conservation and, 68–70
Leases (for homes), 206

LEED certification (*see* Leadership in Energy and Environmental Design certification)
Leisure World, 6
Lesbian communities, 222, 225–226
Levittown, New York, 4
Life cycle costs (of homes), 68
LifeCenter Plus, 41–42, 180–181, 261–266
Lifecenters, 41, 160
LifeSpring Environs, Inc., 64
Lifestyle:
    of boomers, 8, 10–11, 210, 242
    as focus of marketing, 242
    market niches based on, 224–225
    and social environment, 9
    urban, 148
Lifestyle communities:
    age-restricted, 25
    age-targeted, 24–25
    age-qualified, 25
    amenities in, 175–181
    community centers for, 177–178
    market size for, 25–26
    mega communities, 26–30
    multigenerational, 32–35
    smaller-scale, 30–32
Lifestyle directors, 153
Lifestyles of health and sustainability (LOHAS), 17, 224
Lifestyle-targeted condos and apartments, 147–148
Light rail, 47, 48
Light switches, locations of, 80
Lighting:
    in condos and apartments, 156, 159
    energy efficiency of, 70, 71
    in great rooms, 100
    in kitchens, 101
    natural, 90, 156
    technology for, 99–102
    in universal design, 81
    universal design for, 90–92

Lightly Treading, Inc., 68, 71
Linear life paradigm, 7
LinkedIn, 244
Linkletter, Art, 262
Liter, James, 215
Live-work housing, 184
Living rooms:
    in cohousing communities, 112
    design options for, 187–189
    in second homes, 120
Local market research, 169–170
Location:
    of condos and apartments, 148–151
    and market niche, 219–221
    of second homes, 118–120
Lodge, Henry S., 266
Lofts, 41, 148, 159, 217, 242
LOHAS (*see* Lifestyles of health and sustainability)
Los Angeles, California, 42, 160
Lots:
    for one-story vs. two-story homes, 123
    quilting pattern for, 35
    sizes of, 119
Low-E glass, 66
L-shaped kitchens, 190
Lux, Tracy, 238, 239–240
Lynn, Mike, 4, 234

M
Madison, Georgia, 165, 222
Main floor master (MFM) homes, 124 (*See also* One-story living)
Maintenance-free living, 34
    in condos and apartments, 142–144, 159
    in suburban city centers, 151
    in townhouses, 121–122
    in traditional neighborhood developments, 225
    for travelers, 184–185
Malefyt, Timothy, 19
Manor homes, 174
Market area, defining your, 217

Market niches, 215–230
    about, 215–216
    calculating size of, 217–221
    designing for, 227–230
    determining your, 216–217
    planning for shifts in, 229
    types of, 221–227
Market research:
    defining customers with, 235–236
    in the design process, 167–170
    importance of, 215–216
    for market niches, 217
Market size, 25–26, 217
Market studies, 170–172
Market trends (for condos and apartments), 141–143
Marketing, 235–245
    capturing interest with, 235–238
    engagement in, 239–240
    lifestyle as focus of, 242
    messages in, 240–242
    with social networking, 242–245
    timeline for, 238–239
Martha Stewart Center for Living, 208
Martin, Bill, 37
Mass transit, 47–49
Massey, Morris, 9
Master suites, 193, 194, 199, 200
MasterCharge, 10
Materials:
    finish, 181
    recycled, 73–74
McCamant, Kathryn, 55
McCamant and Durrett Architects, 257, 260
McDermott, Maureen, 173
McDonald's, 3
McDowell, Sam, 221
McMansions, 105
Meadowood (community), 52
Mega communities, 26–30
Messages, for boomers, 240–242

MetLife Mature Marketing Institute (MetLife MMI), 271
Mezger, Jeffrey, 203
MFM homes (*see* Main floor master homes)
Michelle Kaufmann Designs, 209
Microblogging, 244–245
Miernowska, Marysia, 56
Migliaccio, John, 21
Millennials, 235
Mindset, of boomers, 8, 235–236
Mobility, 265
Model Code for Accessory Dwelling Units, 205
Model Green Homebuilding Guidelines, 204
Modular building, 135, 209
Monaghan, Tom, 32, 223
Moody Blues, 233
Morning rooms, 198
Motion sensors, 91, 102
Motorcycle insurance, 241
Mt. Sinai Hospital, 208
Mudrooms, 138
Multifamily housing, 36, 45, 49
Multigenerational (mixed-age) communities, 32–35
    age-qualified areas in, 148
    clubhouses for, 154
    cohousing in, 56–57, 251, 259
    fitness centers in, 265–266
    infill communities as, 38
    market niches for, 225
    market research for, 173–174
    single-family homes in, 107
    as trend, 210
Multistory homes, 39
Munksogard (community), 252–254

N
NAHB (*see* National Association of Home Builders)
Nanny cams, 101, 102
National Association of Home Builders (NAHB), 74, 204
Natural features (of land), 227

Natural lighting, 90, 156
Need(s):
    of boomers, 8–9
    of seniors, 249–250
    technology as, 205–206
Neighborhoods, 23–60
    and characteristics of single-family homes, 107, 110–112
    cohousing, 55–60
    considerations of, in the design process, 172–174
    factors in design of, 23–24
    in future of homebuilding, 208–209
    for lifestyle communities, 24–35
    for mixed generations, 35
    rehabs and infill, 36–39
    in suburban city centers, 42–43
    traditional developments, 43–47
    transit-oriented design in, 47–49
    university-affiliated, 49–55
    in urban downtowns, 39–42
Neighbors (in development process), 38–39, 54
Networking (of amenities), 262–263
Networking, social, 24, 209, 242–245
"The New Active Adult Community" (Susan Brecht), 217
New Marketing Network, 8
New urbanism, 4–5, 43, 204–205, 210
Newspaper ads, 237
Niches, architectural, 156, 197
Nielsen's Five Phases, 256
Nifty after Fifty, 180
Nixon, Richard M., 10
Nonskid flooring, 88
The Not So Big House (Sarah Susanka), 136
"Not so big" process, 136–138

Nuclear family, 249
Nursing (in co-care), 252
Nyland Cohousing, 255

O
Oak Hammock (community), 52
Oak Leaf Homes, 54, 173
Obesity, 90
Observatory Village (community), 184
"Offensive design," 74
Office parks, infill housing in, 37
Offices, home, 184, 193, 196–197
One Cherry Lane (community), 37, 177
One-story living, 93, 151, 159
One-way communication, 242
Online communities, marketing to, 243–245
Onsite nursing (in co-care), 252
OPEC, 271
Open spaces (in developments), 227
Orientation, solar, 65–67, 93
Outdoor "rooms," 159, 182, 203
Ovens, 84–85
Owner's manuals (for homes), 73
Ownership, condos and, 146–147

P
Pain points, 138
Painesville, Ohio, 174
Palmer, Arnold, 29
Palo Alto, California, 50
Pantries, 86
Parkhurst, Charles H., 105
Parking:
    in cohousing developments, 56, 248
    for condos and apartments, 142, 147, 158–159
Parks, 179
Parks, Bill, 169–170
Parkview Homes, 30, 173
Parties, marketing with, 240
Past, boomers' views of, 233–234, 241–242

Patience (as sales tool), 239
Patios, 102, 182, 203
PCBC, 271
Pedestrian crossings, 45
Penn State, 51
Perceptible information
    (in universal design), 78
Perryville, Missouri, 163
Personal expression, 133, 183
Peterson, Mary Dene, 60, 260
Pets, 179, 185
Phelps, Jim, 105, 135–136, 203
Phoenix, Arizona:
    apartment developments in,
        159, 217
    loft developments in, 148, 242
    multigenerational communities
        in, 107
Physical effort (in universal
    design), 79
Physical limitations (of boomers),
    76, 77
Physiological enhancement, 262,
    265–266
"The pill," 9
Pines (development), 117–118
Pitt, Brad, 167
POD, 271
Pod system of planning, 35
Pogo, 271
Poinciana, Florida, 28
Political environment (for
    development), 38–39
Pop tops, 128
Porches, 182 (*See also* Entries)
Portland, Oregon, 40, 267–268
Powell, Robert, 218
PreFab Green (Michelle
    Kaufmann), 209
Prefabricated homes, 135, 209
Principles of Universal Design,
    77–79
Privacy, 4, 5, 147
Product sector (in future of home-
    building), 207–208
Professional entities, on social
    networking sites, 243

Projector systems, 96
ProMatura Group LLC, 20,
    113, 168
Protection, 99, 101–102, 120
Protest generation, 4, 7
Psychological reward, 262,
    264–265
Public lands, 119–120, 173
Pulte Homes, 30, 32, 223
Purdue University, 52
Puzzitiello, Richard, Jr., 31

Q
Quail Hollow (community), 37
Quality (of single family homes),
    133–139, 230
Quality committees, 228
Queen Creek, Arizona, 32
Quilting lot patterns, 35

R
Rainbow Vision (community),
    226
Rainwater irrigation systems,
    66, 73
Ramblers, 121
Ranches, defined, 271
Rand, Paul, 187
Real estate market:
    condos and apartments
        in, 141–142
    and recession, 14–15, 19
Rear-loaded garages, 44, 46, 122
Recessed lighting, 91
Recession, 13–21
    about, 13–14
    behaviors and preferences after,
        19–21, 234–235
    as black swan, 202
    boomers' preparedness for,
        17–19
    and boomers' return to
        work, 15
    changes in market from, 19
    and real estate market, 14–15
    and reported wealth of
        boomers, 15–16

    and retirement, 7–8
    and societal shift, 16–17
    and three-generation families,
        203
Recycled content (of materials),
    73–74
Red Quill townhouses, 118, 119
Refrigerators, 84
Rehabilitation (of homes), 36–39
Relationships, marketing using,
    239–240
Religion, 32, 211, 222, 223, 266
Remodeling, 36, 209
Renters:
    affluent, 144–146
    demographics of, 142
    of second homes, 120
Resort-styled communities,
    110–111, 131
Retirement:
    consumer confidence about,
        13–14
    psychological reward after,
        264–265
    and recession, 7–8
    saving for, 18
    working after, 7, 15–17,
        183–184
Retirement Confidence Studies,
    13, 15
Reverse second homes, 42,
    120–121
Rhoad, John D., Jr., 150, 152–154
*Right House, Right Place, Right
    Time* (Margaret Wylde),
    64, 75
Rightsizing, 137–139
Rightsizing Your Life (Ciji Ware),
    137, 147
Riverbend Apartments, 144, 145
Rivers Bend (community),
    179–180
RMJ Development Group
    LLC, 150
Rock and Roll, 9
Rock Creek Park, 225
Roll-in showers, 81, 89, 195, 196

Roman Catholic Church, 32
Romero, Roccio, 135
Rosetta, Pennsylvania, 256
Rouse, James W., 163
Rouse Company, 163
Row houses, 121–122
Rural villages, 120

S
Sagan, Carl, 95
San Clemente, California, 31
San Francisco, California, 40
Sandwich generation, 125, 226
Santa Cruz, California, 129, 226
Santa Fe, New Mexico, 226
Satisfaction surveys
    (for buyers), 236
Schaumburg, Illinois, 42
Schreiner, David, 18–20, 40
Screened porches, 182
Seats (for bathtubs), 89
Seattle, Washington, 40
Second homes:
    market niches for, 224
    reverse, 42, 120–121
    as single-family homes,
        117–121
Security:
    in condos and apartments, 151
    for second homes, 120
    technology for, 99, 101, 102
See-through floor plans, 27, 28,
    198–199
Senior cohousing, 56, 57–58,
    250, 251
Senior Cohousing (Charles
    Durrett), 55, 225
Senior Cohousing Handbook,
    2nd Edition (Charles
    Durrett), 247
Seniors:
    boomers vs., 6, 234
    designing for, 2
    needs of, 249–250
Sense of community, 256
Service sector (in future of home-
    building), 207–208

Sexual preference, market niches
    based on, 222, 225–226
Shanghai, China, 48
Shelves, adjustable, 86
Showers:
    roll-in, 81, 195, 196
    televisions in, 102
    universal design for, 88–89
    and water conservation, 73
Sidekick Homes, 226
Silent generation, 6–8, 18, 234
Silver, Joy, 225
Silver Sage (community), 57–60,
    223, 249, 254–255, 257
Simplicity (in design), 134–135
Simplicity of use (in universal
    design), 78
Singapore, 48
Single-family homes, 105–139
    accessory dwelling units,
        124–130
    architects' preconceptions
        about, 113–116
    and cohousing, 249
    duplexes and triplexes, 121
    features of, 112–113, 130–133
    neighborhood-driven character-
        istics of, 107, 110–112
    quality and quantity in,
        133–139
    second or vacation homes,
        117–121
    sizes of, 105–109
    townhouses or row houses,
        121–122
    two-story, 122–124
Single-story homes, two-story vs.,
    123–124
Sinks, 85, 89, 195
Site placement, 181
Sites, 66, 119
Size and space for approach and
    use (in universal design), 79
Small towns, 221–222
Smart Growth movement,
    204–205
Smoke detectors, 101

Social gatherings, marketing
    with, 240
Social networking, 24, 209,
    242–245
Social niches, 222–224
Socialization, 262, 263
Sol Vista Ski and Golf
    Resort, 117
Solar orientation, 65–67, 93
Solar photovoltaic arrays, 72
Solar Row (community), 257
Solera at Johnson Ranch
    (community), 32, 223
Solivita (community), 28–29, 180
South Lebanon, Ohio, 173
South-facing elevations, 67
Space age, 9
Spirituality, 211, 262, 266
    communities based on, 32, 60
    market niches based on,
        222, 223
Spousal loss, 263
Sputnik I, 9
Stahr, Rebecca, 64
Stairs, 81, 91, 197
Standard Pacific Homes, 31
Stapleton (community), 35
Stemen, Mark D., 148, 152–155
Stern, Robert A. M., 149
Stewart, Martha, 208
Storage:
    in condos and apartments, 147,
        158–160
    design options for, 197–198
    in kitchens, 86–87
    for travelers, 184–185
Storm-resistant design, 209
Stoves, 84
Streets:
    in cohousing, 56, 248
    in traditional neighborhood
        developments, 44, 46
Study Groups (for cohousing),
    250, 255, 256
Suburban city centers:
    condos and apartments in, 143,
        161–165

maintenance-free living
in, 151
neighborhoods in, 42–43
wood frame construction
in, 162
Suburbs, 131, 249
Summer kitchens, 196–197
Sun City, 6, 110, 218
Sun City Huntley (community),
27, 171–172
Sun Valley (community),
117–118
Surveys:
of buyer preference, 168–169
of customer satisfaction, 236
Susanka, Sarah, 136, 137
Sustainability, 63 (*See also* Green
building design)
Swimming pools, 110
for apartments and condos, 154
in development design, 179
in lifestyle communities, 30, 31
in traditional neighborhood
developments, 44, 46

T
Taleb, Nassim Nicholas, 202, 209
Talega Gallery (community),
31–32
Task lighting, 91–92
Technology, 95–103
Boomers' comfort with, 95–97
as builder opportunity, 98
buyer expectations of, 97–98
for control in home, 98–99
in future of homebuilding,
205–206
installation process for,
100–103
seniors views of, 234
Television, 3, 100, 102, 245
Tenenbaum, Louis, 81, 88, 99
Three-generation families,
125–127, 203
Timeline (for marketing),
238–239

Time-shares, 206
TNDs (*see* Traditional neighborhood developments)
Toilets:
comfort-height, 270
compartments for, 87, 90, 195
dual-flush, 73
Tolerance for error (in universal
design), 78–79
Towel racks (as grab bars), 88
Town centers:
condos and apartments in,
163–165
and suburban city centers,
42–43
in traditional neighborhood
developments, 44, 45
Townhouses:
bedroom location in, 199
as single-family homes,
121–122
in traditional neighborhood
developments, 45
in transit-oriented designs, 49
Traditional neighborhood developments (TNDs):
market niches for, 225
as neighborhood type, 43–47
single-family homes in, 111–112
views in, 131
Trailing-edge boomers (late
boomers), 1, 6
Trails (*see* Walking paths and
trails)
Trailview Development
Corporation, 60
Trains, 47, 48
Transfer benches (for bathtubs),
89, 195
Transit-oriented design (TOD),
47–49, 222
Travel, 184–185, 234
Trends, 201–212
accessory dwelling units,
204–205
community types, 208–209

and diversity of boomers, 210
and forces of change, 211–212
green building design, 204
home sizes, 203
identifying, 201–202, 209
outdoor "rooms," 203
product and service sectors,
207–208
in remodeling, 209–210
and technology, 205–206
three-generation families, 203
uses of homes, 206
Triplexes, 121
Tub seats, 89, 195
Turkey, 191, 192
Twain, Mark, 269
Twitter, 24, 244–245
Two-story single-family homes:
bedroom location in, 199, 200
costs of, 131, 132
elevators for, 49
for infill sites, 39
one-story homes vs., 122–124
in traditional neighborhood
developments, 39
Tyson's Corner (community), 228

U
UARCs (*see* University-affiliated
retirement communities)
UBRCs (university-based
retirement communities), 50
UD (*see* Universal design)
ULRCs (university-linked retirement communities), 50
Unitarian Universalist Church, 60
United States Green Building
Council (USGBC), 272
Universal design (UD), 63–64,
74–92
and aging in place, 74–75
for apartments and condos,
156–157
for bathrooms, 87–90
for entries, doors, and hallways,
81–82

and green building design, 63–64
guidelines for, 80–81
homeowner's knowledge of, 75–77
interest in, 92–94
for kitchens, 82–87
levels of, 79–80
lighting in, 90–92
marketing with, 241
principles of, 77–79
as trend, 210
windows in, 82
University of Florida, 52
University of Montana at Missoula, 54
University of Notre Dame, 52, 207
University Place (community), 52, 207
University-affiliated retirement communities (UARCs), 49–55
classes in, 184
current, 52–53
guidelines for, 55
healthcare components of, 207
and location-based market niches, 222
market niches for, 226
mutual benefits of, 50–52
new generation of, 53–54
University-based retirement communities (UBRCs), 50
University-linked retirement communities (ULRCs), 50
University-related retirement communities (URRCs), 50
Urban areas (downtowns):
condos and apartments in, 143, 148–150, 160–161
and location-based market niches, 221
neighborhoods in, 39–42
wood frame construction in, 160

URRCs (university-related retirement communities), 50
U.S. Department of Housing and Urban Development (HUD), 25, 130, 204–205
USGBC (United States Green Building Council), 272
U-shaped kitchens, 190

V
Vacation homes, 117–121
Vail, Colorado, 118
Vallagio Condominiums, 144–146
Values (of boomers), 133–134
Vanicessity, 4, 234
Vanities, bathroom, 89–90
Vaporware, 205
Vastu Shastra, 191, 202
Venetian Falls (community), 176, 178
Venice, Florida, 178
Verona, Wisconsin, 119
Vestibules, 120
Vietnam War, 6, 9
Views (from homes), 131, 181–182, 198–199
Villa Italia mall, 162
The Village at Penn State, 51
Village greens, 44, 46
Villages, 120, 173
The Villages (community), 26, 29–30, 173, 180
Villas, 121
Vision, community, 24, 172, 228–229
Visitability (in universal design), 79
Visitors, 137
Visual elements (of architecture), 181–183
Voice recognition software, 96

W
Walkable streets, 46
Walk-in bathtubs, 89, 195

Walking paths and trails:
creating, 168, 178–180
in university-affiliated communities, 51, 180
Walking-oriented communities, 225
Wallace, Priscilla, 8, 206, 210, 212, 234, 241
WALT (Wellness Assisted Living of Texas), 208–209
Ware, Ciji, 137, 138, 147
Washington, D.C., 40
Waste, controlling, 73–74
Wasted space, 139, 155
Water conservation, 66, 72
Watergate scandal, 10
Wealth (of boomers), 15–16
Welch, David, 19, 21
Wellness Assisted Living of Texas (WALT), 208–209
West, Mae, 233
West Lafayette, Indiana, 52, 207
White Fence Farm (community), 228
Wii, 34
Wild Sage (community), 58, 257, 259
Windows, 66, 82, 90
Winter Park, Colorado, 118
Wolf Creek Lodge (community), 60
Wonderland Hill Development Company, 58, 59, 226, 247, 255, 257–259
Wood frame construction, 160, 162
Woodlands (community), 228
Woodstock, 10, 233
Word-of-mouth advertising, 237
Workforce, returning to, 7, 15–17, 183–184
Workforce housing, 118
Workspace:
in homes, 136–137, 183–184, 196–197
in kitchens, 85–86

World War II, 1, 4, 234
Wortham, Willie, 180
Wright, Steven, 141
Wylde, Margaret, 21, 46, 64, 68,
    75, 113, 131, 135, 136, 142,
    146, 150, 152, 158, 168,
    169, 175, 187, 189, 191,
    193, 195, 197, 198, 228,
    230, 235, 236

X
Xeriscaping, 73

Y
YMCA, 180
Younger Next Year (Henry S.
    Lodge and Chris Crowley),
    266
YouTube, 245

Z
Zero-step (no-step) entries, 76,
    80, 81, 93, 110
Zolli, Andrew, 16, 17, 212
Zoning rules:
    about clustering, 173
    for accessory dwelling units,
        205, 267–268
    for age-restricted housing, 21